AN ILLUSTRATED HISTORY OF
CIVIL ENGINEERING

AN ILLUSTRATED HISTORY OF **Civil Engineering**

J. P. M. PANNELL

M.B.E., M.I.C.E., M.I.MECH.E.

228 black-and-white illustrations

THAMES AND HUDSON · LONDON

Contents

FRONTISPIECE

Temporary works, *such as the centring used in building the Ballochmyle Viaduct on the Glasgow and South-Western Railway, are as much a part of the civil engineer's art as the finished structure itself.*

Foreword

IT HAS BEEN SAID that all progress is but a continuation of history, and it is very certain that many modern advances in materials and method are, in essence, a repetition of things which have been done before, although the detail may vary.

This point was brought home to me very forcibly as I read the chapters of Mr Pannell's book.

There is much of absorbing interest to record in the history of engineering, because its achievements are, in a very real sense, the physical basis of modern civilization, but unlike other aspects —politics, philosophy, and the arts—it has not until now been adequately collated, and even less adequately recorded in historical sequence. This is a great pity, because the history of man's efforts in forming and improving his physical environment is a fascinating story and is shown by this book to be also a most useful background for future progress. Historians in this field are very few, but the research which obviously has occupied the author for a number of years entitles him to stand alongside Samuel Smiles, and his efforts will undoubtedly be appreciated in interest, as well as in usefulness, by all who read his book, which is written in a manner certain to appeal to the layman as well as those who choose engineering as a career.

SIR HERBERT J. MANZONI, C.B.E., D.SC., M.I.C.E.
Past President of The Institution of Civil Engineers
City Engineer of Birmingham

Date	Roads	Rivers and Canals	Railways	Docks and Harbours
0	Trackways	Egyptian canals First Suez Canal	Grooved carriageways	Mediterranean ports Pharos of Alexandria
	Itinerarium Antonini Saxon roads	Fossdyke		Roman harbours
1000	King's Highway	Naviglio Grande Invention of pound locks		Timber and stone quays Portsmouth dry dock
1500	Highways Act	River improvements Exeter Canal First dredger	Tramways in mines	Hanseatic League
1650	Turnpike Acts begin	Canal du Midi	Horse tramroads	Winstanley's Eddystone lighthouse
1700	General Wade's roads	Continental canal expansion		Howland Great Wet Dock
1750	Tresaguet's French roads	Sankey Canal Bridgewater Canal Pontcysyllte Aqueduct	Iron rails	Smeaton's Eddystone lighthouse
1800	Holyhead road McAdam	Caledonian Canal Berkeley–Gloucester Canal	'Rocket' locomotive Stockton and Darlington Railway Railway boom	Plymouth breakwater St Katherine's Docks, London Southampton Docks begun
1850	First concrete roads Asphalt surfacing	Suez Canal Corinth Canal Manchester Ship Canal	First transcontinent lines Mont Cenis tunnel Tube railways	Tilbury Docks
1900	Motor transport Mass production of cars	Kiel Canal Panama Canal	Trans-Siberian Railway Electrification	Dover Harbour Port of London Authority
1920	Continental motorways	Tennessee Valley Authority Volga–Don Canal	Railway amalgamation	Zuyder Zee dyke Tilbury Docks extension
1940	British motorways Traffic engineering	St Lawrence Seaway	Nationalization in Britain	Mulberry Harbour Dutch Delta plan Rochdale Report

Water and Health	Bridges	Materials	Structural Theory	Construction Methods
Water and sanitation at Knossos Pont du Gard	Corbelled arch Pontoon bridges True arch	Brickwork Masonry Metals	Column and lintel Greek scientists	Mass use of labour Surveying Water power
Roman sanitation	Trajan's Danube bridge		Vitruvius' *De Architectura*	Cranes Coffer-dams Contracts
Ecclesiastical water supplies	Pont St Bénézet London Bridge Ponte Sisto, Rome		Leonardo da Vinci	Craft guilds
Drake's Leat The 'New River'	Ponte di San Trinita, Florence	Charcoal iron	Galileo's 'Two New Sciences'	Horse gin Treadmill Capstan
Versailles waterworks York Buildings waterworks	French influence on bridge design		Hooke's Law Newton's *Principia*	2,000-ton blocks handled at Tangier
Steam pumping Cast-iron water mains	Gautier's *Treité des Ponts*	Iron smelted by coke Tensile testing machines	Arch theory Principle of virtual displacements Euler's strut theory	Diving bell Clay puddling
Water closets	Edwards' Pontypridd Bridge Coalbrookdale Iron Bridge	Coulomb on soil testing Hydraulic cement	Coulomb's researches Girard's 'Strength of Materials'	
Simpson's filter beds Chadwick's sanitary report Dr Snow investigates cholera	Chain suspension bridge in U.S. Captain Brown's flat link chain Menai Bridge	Cort's inventions in iron-making Portland cement First rolled sections	Navier's equations of elasticity Truss design	Tunnelling shield Compressed air Steam power machines
Liverpool gets water from Wales	Saltash Bridge Brooklyn Bridge Forth railway bridge	Bessemer steel Metal fatigue studied Reinforced concrete	Continuous beam theory Strain energy equations	Hydraulic power tools Steam excavators
Aswan Dam Croton Dam Hydro-electric development		Electric welding Aluminium Pre-stressed concrete		Tracked vehicles Internal combustion engines Concrete mixers
Arch dams	Sydney Harbour Bridge Golden Gate Bridge	Special cements Vibrated concrete Alloy steel structures	Photo-elasticity Moment distribution Relaxation methods	Air-placed concrete Steel forms
Pre-stressed concrete pipes Large-scale distillation Australian Snowy Mountain scheme	Steel centring	Plastics Bonded laminated timber	Soil mechanics Electronic computation Ultimate load design	Prefabrication Centralized concrete making Bulk electricity supply

Introduction

CIVIL ENGINEERING is a branch of human activity which has been pursued from those remote times when man began to adapt his environment to his needs. As with the products of the potter and the worker in bronze, the works of the civil engineer, possess a permanence which has enabled us to recover, at least in part, some examples of his skill dating from many thousands of years ago.

The history of civil engineering is a subject worthy of study, and almost every branch of the profession has a wealth of unrecorded material, awaiting the interest of the student. This almost untapped reservoir lies in plan stores, minute books, account books, contract documents, and many other written and drawn records. Even more important, it is still in existence as part of the equipment of our daily life in the form of roads, aqueducts, canals, bridges, etc., each the achievement of someone, perhaps long forgotten, but often worth the labour of enquiry into his life and times.

The activities of the civil engineer are entirely beneficent. The description 'civil engineer' arose from the need to distinguish engineering work performed for the public good from that of a purely military nature and, when the Institution of Civil Engineers which was established in 1818 applied for its charter in 1828, the scope of civil engineering at that time was clearly defined by Thomas Tredgold in words which, like those of any other classic, though often quoted, will bear any amount of repetition.

'That species of knowledge which constitutes the profession of a Civil Engineer; being the art of directing the great sources of powers in Nature for the use and convenience of man, as the means of production and of traffic in States both for external and internal trade, as applied in the construction of roads, bridges, aqueducts, canals, river navigation and docks, for internal intercourse and exchange; and in the construction of ports, harbours, moles, breakwaters and lighthouses, and in the art of navigation by artificial power for the purposes of commerce; and in the construction and adaptation of machinery; and in the drainage of cities and towns.'

Although other, and more specialized, institutions have since arisen, which tend to concentrate on some parts of the scope of the civil engineer as defined in these words, the 'Civils' still open their doors to all engineers of appropriate attainments in any of the branches of engineering referred to in Tredgold's definition. Indeed, practitioners in many new applications of engineering, unthought of in his time, are today included in the Civils' membership.

The very permanence of their works makes civil engineers close followers of tradition, and the great responsibility for public safety which lies on him makes every member of the profession cautious in his approach to innovation. This does not mean that progress is inhibited, but rather that every device for the advancement of knowledge of his subject is used by the designer of great works before he commits himself to a final solution of any problem. The great engineers of the eighteenth and nineteenth centuries excelled in this respect and, through their influence, the professional approach has been one of decision based on the fullest possible information.

From the time of John Smeaton onwards, the traceable links between the great men in civil engineering and their successors are very strong. These links may be followed backward, by a great number of engineers practising today in most countries of the world. It can be a salutary exercise for any young engineer to trace back such influences as far as they apply to himself and to find, by so doing, how much he owes to men often forgotten and ignored by those who, under the pressure of events today, overlook the significant contributions made to their knowledge by their professional ancestors.

Pursuing further the concept of the civil engineer being a public benefactor, a closer look at his function shows that a great part is devoted to improvements in communications and public health. Many a youthful entrant to the profession aspires to be a great builder of bridges and during his lifetime he may achieve this object, but on the way he will probably be engaged in the construction of roads, drainage works, waterworks, docks, plant for the disposal of sewage, or foundations and culverts for a great power station. Many of the projects on which he will be occupied will, in time, change the pattern of life of whole communities and enable increasing numbers to enjoy a higher standard of living. Any doubts in this respect will be dispelled by a study of the pattern of the past, for the history of civil engineering makes it abundantly clear that the work of the profession leads to the majority of people enjoying ever rising levels of material well-being. Whether the moral and spiritual standards of mankind progress at the same rate is a question beyond the scope of this book, but it can at least be said that the ethical standards maintained in the civil engineering profession, following the example of its early exponents, are as high as those to be found in any walk of life.

This book stems from courses of lectures given by the author to second-year civil engineering students at the University of Southampton during the sessions 1959–60 and onwards. Although there is no lack of material on which to base such a course, there is no comprehensive work of British origin covering the whole subject. This book, while not entirely filling the gap, is intended to provide an outline of as much of the subject as can be covered in a readable form within the scope of its size. Inevitably, a book of 70,000 words or so, if it is not to be a mere list of events and people, must suffer omissions to gain space for longer descriptions of selected examples, even at the risk of criticism in respect of the parts left out. The author has taken that risk. He has also chosen to describe more fully those parts of history which have come within his own personal experience or contacts in the belief that his writing may thereby be enlivened for the reader's benefit.

The sources cover a wide range, and reference to every one would be difficult; those recommended for further reading are given in a separate bibliography. Some acknowledgment must, however, be made to such classics as Smiles' *Lives of the Engineers* and his *Industrial Biography*, Cresy's *Cyclopaedia of Civil Engineering*, and Telford's *Autobiography*. Of the modern authors, Charles Hadfield on Canals and L. T. C. Rolt, with his unapproachable standard of writing in his trilogy on Brunel, Telford and the Stephensons, deservedly enjoy a high reputation. The Presidential Addresses of the Institution of Civil Engineers include many dealing in the history of branches of civil engineering, and the papers of the Newcomen Society are also a valuable source.

No book of this kind can be written without the co-operation of many people. The author gratefully acknowledges the help received. Professor P. B. Morice suggested that it should be written and has given his constant help in its preparation, especially in respect of Chapter VIII. Thanks are due to the Librarians and their staffs at Southampton Public Library, Southampton University, Bristol University, The Institutions of Civil and Mechanical Engineers, Mr P. S. Peberdy, the Curator of Museums and Mr B. C. Jones, the Archivist of the Borough of Southampton. The author's many friends in the civil engineering profession have not only provided material and suffered him to dip into their archives, but have, by reading and criticism of the various chapters, actively assisted in the making of the book. For her work in typing, the author gratefully acknowledges Miss D. Angel's meticulous care. He also greatly appreciates the care and expert knowledge put into the preparation of this book by the publishers and, last but not least, once again pays tribute to the patience and encouragement displayed by his wife throughout the period of its preparation.

Bronze and Iron Age roads in Britain — the Roman
road system and the Appian Way — classes of
Roman roads — Roman roads in occupied Britain
the tradition of the 'King's Highway'
maintenance of roads by the Church — Turnpike Acts
General Wade's military roads in Scotland
'Blind Jack' Metcalfe of Knaresborough
Telford's London–Holyhead road — McAdam's
road-making — the first concrete roads
history in roads

◄The first great road-builders *in Britain were the Romans who needed*
to link their garrison towns. This stretch of Watling Street, near Little
Brickhill, Buckinghamshire, is still in use today.

Roads

THE CONSTRUCTION AND maintenance of roads has probably involved more time and money than any other branch of civil
engineering, but paradoxically, the earliest roads were not
engineered. They developed as tracks and were the easiest route
for travelling from one place to another.

In Britain, the road system dates back more than 3,000 years,
the network of the Bronze and Iron Ages being still traceable.
The country at that time was almost entirely covered with forest,
and only the high uplands were free from thick undergrowth.
The ridges, or watersheds, thus became the trade routes, and along
them for centuries an interchange of trade and ideas took place.
Along these roads settlements developed; man fought, bred
cattle, worked with tools and worshipped his gods. The road
intersections became places of importance and around them
leaders of thought and religion tended to congregate; thus at
such intersections we find great monuments like Stonehenge and
Avebury. Around these places great men were buried under
mounds which, like the monuments, were examples of civil
engineering. Their settlements were surrounded by earthworks
for defence and for the security of their beasts. Only by even
greater civil engineering works could the traces of the lives of
these people be removed from the earth and thus the ridgeways—
the Harrow Way, the Lun Way, the Fosse Way—remain to tell
us of those early road users.

The Romans, in the expansion of their Empire, were the first
great road engineers. To them, engineering was a vocation leading to high office in the state and one which competed with
military art for the attention of the rulers. The military importance of roads led to high standards of construction and maintenance, even in the outer fringes of the Empire. The pattern of

military occupation of countries like Britain consisted largely of garrison towns, built and run on Roman lines, connected by a highly organized system of communications depending mainly, but not entirely, on roads.

The main Roman road system took the form of twenty-nine great military roads centred in Rome. The Roman Empire extended over eleven regions—Italy, Spain, Gaul, Britain, Illyria, Thrace, Asia Minor, Pontus, the East, Egypt and Africa. These regions were divided into 113 provinces traversed by 372 great roads totalling 52,964 Roman miles. The greatest of these roads as a piece of engineering must have been the Appian Way. This extended for 360 miles of which 120 miles was paved with large slabs of stone each squared and fitted exactly; the remainder was paved with fitted polygons of lava. Commenced by Appius Claudius after whom it was named, and who built 142 miles, it is believed to have been finished by Julius Caesar.

The largest prehistoric earthwork *in Britain is Silbury Hill, near Avebury in Wiltshire. A Roman road passes close by it.*

The Via Appia *was the greatest of all the Roman roads. It extended for 360 miles and the width was about 14 ft, sufficient for the passage of two carriages.*

The Romans even had the equivalent of the modern road-book, the *Itinerarium Antonini*, compiled about A.D. 200. This itinerary set out the military stations on the principal roads and the distance in Roman miles along them. A Roman mile was a thousand paces (*milia passum*), a *passus* being measured by the position of the rear foot at the beginning of a pace to the position of the same foot after taking it, a distance slightly under 5 feet, making the Roman mile approximately equal to or a little less than an English mile.

Roman roads were engineered to fulfil a purpose and may be divided roughly into five classes: *Via*—a common road for the passage of two carriages without collision, width about 14 feet; *Actus*—a road for a single carriage deriving its name from a measure of land about 7 feet by 200 feet; *Iter*—a road for pedestrians and horsemen five feet wide; *Semita*—half the width of the *Iter*; *Callais*—a mountain road for attending flocks.

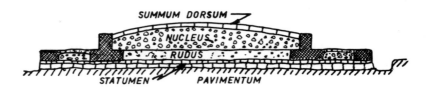

The five layers *of the Roman road raised the surface well above the surrounding ground level, forming a causeway with a cambered surface. By such means the Romans provided roads which have lasted for 2,000 years and which today are the foundations of modern highways.*

The Romans constructed their roads in several layers, building them up to a causeway, thus achieving the drainage necessary for all the year round use. Before a road of the highest class was made, two parallel trenches were cut along the intended edges of the formation and, between these, all the top soil and loose material was removed down to hard ground. This excavated cut was then filled in with fine, dry earth well rammed in, forming a layer called the *pavimentum*. Next, a layer of small, squared stones, either left dry or with mortar poured in to form a watertight course, called the *statumen*. On this was laid the *rudus*, or *ruderatio*, a kind of concrete consisting of one part of broken stones to two parts of lime. The *nucleus* was the next layer and was a lower grade of material made of lime, chalk and broken tiles, or gravel, or a mixture of sand, lime and clay; in fact, whatever local material could be made to consolidate into a hard, permanent mass (for instance the debris and slag from iron-working areas on the London–Lewes Way which ran from Peckham to the South

The causeway or 'agger' *is revealed by the excavation for this modern road-widening scheme on Ermine Street, between Cirencester and Birdlip.*

Downs). This Roman road was probably built to open up the iron-smelting sites of Sussex. Finally, the wearing surface, or pavement, called the *summum dorsum* or *summa crusta* consisting of fitted stones like flagstones, rectangular or polygonal in shape, or, in the case of roads of less consequence, a wearing surface of a gravel and lime concrete. These layers raised the road surface well above the surrounding ground level, forming a causeway known as the *agger*.

Great care was taken to keep water out of this formation, by cambering the top surface and by cutting drainage ditches along the edges of the road. By such means the Romans provided roads which, even in countries subject to wide variations of weather, have lasted 2,000 years and even now are, for many hundreds of miles, the foundations of modern highways carrying traffic far beyond anything the builders could have imagined.

Much is already known of the Roman road system in Britain; the Ordnance Survey map of Roman Britain portrays an im-

A system of military roads *in
Britain was established by the
Romans during their four centuries
of occupation. It was based on a
pattern of four main road groups
radiating from London and a fifth
running across the country
between Exeter and Lincoln.*

The Ridge Way *is crossed at
Chisledon by the Roman road to
Swindon and Gloucester.*

pressive network of roads identified with certainty and indicates
the probable course of many others. Great progress has been made
during recent years in discovering extensive stretches of the
road system, probably the greatest contribution to this work
being made by aerial photography. Initiated by O. G. S. Craw-
ford, it has since been the principal method of tracing not only
the roads, but other works of early man of all ages. Air photo-
graphs, to be of full value, need to be followed up by investigation at and below ground level, and in this respect much remains
to be done. Strangely enough, the lengths of Roman road hardest
to identify are often those which have been in use continuously
since they were built. In such cases the original formation has been
cut into and reconstructed so many times that little or no certain
traces remain. Where, however, the old Roman road connected
places which ceased to be of importance after their day, traffic
ceased. The road formation, preserved perhaps under layers of
mould, remains unharmed. Such a road is Sarn Helen, the ancient
highway running between North and South Wales. The origins
of this road became lost in legend and in the Mabinogion the

22

construction was attributed to the influence of Helen Luyddawc, a British maiden. It is, however, more probable that the name is a corruption of Sarn y Lleng—the Road of the Legions—providing as it did, communication for military purposes and for trade routes between the areas of mineral wealth in Wales then being exploited for the central Roman economy.

All the roads ultimately led to Rome, the centre of the system, where eventually the civilization became entirely dependent on tribute from the occupied Empire not only for goods, but for man-power. Not only Rome, but other great civilizations were extravagant in the use of human life. Slave labour was plentiful and easily replaced; on works of great magnitude the mortality rate was high, there was no 'Factories Act' in those days. This dependence on conscripted labour, which extended to the military forces, and its consequent ease and softness of living, led to the decadence and ultimate downfall of the Roman Empire and the withdrawal of its influence in the occupied countries.

So great, however, had that influence been, that for many hundreds of years the less sophisticated successors of the Romans were able to draw on the reserves of resources laid down during the occupation. Thus, the road system of the Romans provided better communications for Saxon, Danish and Norman invaders than any in later times until the eighteenth century. These people used the roads, not knowing their origin, and gave them names which remain to this day—Stane Street, Watling Street, Akeman Street, Ermine Street and Fosse Way. Down the Roman road rushed Harold to meet the Normans at Hastings. A forced march such as his could not have been achieved without a well-known and marked military road, built nearly a thousand years earlier and probably not maintained after the builders left the country to its fate.

During that period Britain ceased to be a coherent whole. Such government as there was tended to be more local and trade over long distances died out. Small communities became self-sufficient and during that time a secondary road system was born which became the basis of the local roads of our own time. These roads were not engineered roads but merely highways over which pack animals, cattle and pedestrians could go from place to place. The constant use of these highways over the years led to wear. Mud churned up during the winter became dust under feet and hooves in summer and was blown away. Such routes became the lanes or hollow ways characteristic of some parts of the country, particularly in Devon and Cornwall. These hollow ways, of their very nature, became natural land boundaries and their alignment has been preserved to the present day, in many cases perpetuating the original artificial land divisions past which the traffic made its way.

After the Norman Conquest, a system of national communications became necessary for military purposes. Even before that time, certain roads had assumed greater importance for military or trade purposes; the word *street* derives from this concept and this developed into the tradition of the King's Highway. In the late eleventh century the need for through routes of travel led to the adoption of four great roads, based on the Roman system, Watling Street, Ermine Street, the Fosse Way, and the Icknield Way. On these roads travellers were protected by 'the King's peace', so that they could travel without molestation on their own affairs or those of the king. Protected by the law, travel on these roads became the model of travel generally and the idea of the King's peace extended to the system of national highways throughout England and Wales.

This encouragement to travel brought great benefits to the country; not only was it possible for the King's justice to be dispensed impartially by travelling courts, but trade spread widely across the land and, with finance focusing on London, encouraged great extensions of trade with the Continent, and, later, the Near

Fosse Way was one of the four great roads in Britain based on the Roman system.

East. The conjunction of this road system with navigable waterways led to the establishment and growth of many of our modern towns so that by the twelfth and thirteenth centuries it may be said that the national geographic pattern was established and recognizable for that which we know today.

With the Normans came a great spread of organized religion and the establishment in England and Wales of the great ecclesiastical houses. These attached great importance to communications and became very concerned over the neglect of the country's road system. The expression of opinion by religious leaders and their influence on public activities led to the maintenance of roads and bridges being considered as work of piety. This led, through the issue of indulgences and the giving of benefactions, to the provision and administration of sufficient funds for the maintenance of road communications between centres of religious importance, affecting the whole country.

As the building programme of the Church virtually organized the supply of skilled masons, carpenters and other constructional trades, the administration of roads and bridges was effective, and until the dissolution of the monastic houses by Henry VIII the road system of the country was maintained in reasonably adequate condition for the means of transport of the period. All movement was by horse or foot and the requirement was therefore based on the provision of crossing places at rivers and on an adequate width of highway to allow for the avoidance of bad places in wet weather. Main roads not only required the width for traffic to change its route to avoid quagmires and potholes, but to enable the traveller to keep clear of hedges or any cover where thieves might be lurking. The adequate width available for many of our modern road improvements dates back to those days when width was a safeguard against robbers.

Bad though they may have been, the roads deteriorated still further when deprived of the interest of the Church. In Tudor and Stuart times increasing prosperity brought by our overseas successes might have brought greater benefits to the country if the internal communications had not been so atrociously bad. Attempts were made in Tudor times and later under Stuart rule to impose the responsibility for road maintenance on to the parishes. This responsibility had always existed under common law and an Act of Parliament was passed in 1555 endeavouring to strengthen the position. The Act provided 'For amending the High-ways being now both very noisom and tedious to travel in and dangerous to all Passengers and Carriages'. This title indicates that wheeled vehicles had by then become an important part of communications.

The 1555 Act provided for the election by every parish of two parishioners as Surveyors and Orderers to take office for one year or the payment of a fine on default. Their duty was to nominate four days (later increased to six) between Easter and Midsummer on which 'the Parishioners shall endeavour themselves to the amending of the said Ways'. Wingate, in his *Abridgement of Statutes* (1675) says:

> 'The Officers and days being thus appointed, everyone having a term, or Ploughland either in arable or pasture, is chargable to send two able men with a team and tools convenient to work 8 hours upon every one of those four days, in pain to forfeit 10s. for every day that default is made . . . And every Cottager is bound to work himself, or to find a sufficient labourer to work for him, as aforesaid, in pain to forfeit 12d. for every day that either of them makes default.'

These provisions failed, largely because human nature is what it is. The parishioners tended to nominate the most unpopular and possibly incompetent men in the parish; they cared little about the value of their stretch of road as a part of the national economy, and, even more important, the labour and equipment were conscripted during the busiest season on the farms. Wages were low and savings non-existent, so that the labourers tended to spend the conscripted time in begging from the passers-by. The worst roads were in districts where stone was scarce or where there was none, so that the problem of materials faced the Surveyor. This was often solved by robbing the works of previous centuries and resulted in the destruction of prehistoric monuments, monastic ruins and the Roman roads where those sources existed nearby.

Many writings of the seventeenth century describe the state of the roads of the time and it is evident that as travel increased in volume, the neglect of the roads led to them reaching a worse state than had ever existed before. This led, in 1663, to the passing of the first Turnpike Act, 'For Repairing the High-ways within the Counties of *Hertford, Cambridge* and *Huntington*'. This Act established the new principle of making the road *users* pay for the maintenance of the road by extracting from them a tax or toll each time the road was used. Although this was a very practical solution to the problem, the imposition of a charge on road use led to many violent complaints and the destruction of the toll bars. In the case of the first turnpike road, three gates were named in the Act; 'for the county of Hertford at WADES' MILL; and for the county of Cambridge at CAXTON and for the county of Huntington at STILTON'. Of these three gates, only one succeeded in fulfilling its function, the one at Wades' Mill. The Stilton gate was never erected as local opposition was so violent, and the Caxton gate was so easily evaded that it was ineffective.

Eventually, however, reason triumphed and by 1848 when Lord Macauley wrote his *History of England* 30,000 miles of turnpike road had been established in Britain.

Turnpike legislation proceeded slowly at first so that by 1750 there were 169 Acts in existence; the next twenty years brought this number up to 530 and from that time the number increased to over 1,100 by the 1830's.

At first Turnpike Acts were limited to twenty-one years' duration on the assumption that tolls might be sufficient to clear the cost of a road in that time. In practice, however, this rarely or never happened and in 1830 the term was increased to thirty-one years, by which time the total debt of the turnpike trusts amounted to £8,500,000, of which £1,000,000 was unpaid interest. Bond debts of the turnpikes in 1830 amounted to over £7,000,000 on which £300,000 interest was paid annually. Out of an annual income of £1,800,000, road works cost £1,064,000 and management £135,000.

Company of Proprietors
of
BURSLEDON BRIDGE and ROADS.

Nº 52

These are to certify that Isaac Galpin of the Town
and County of the Town of Southampton Gentleman is a Proprietor
of One One hundred & fifty fifth Share of and in the
Bursledon Bridge and Roads Undertaking
Subject to the Rules Regulations and Orders of the
Company of Proprietors of the said Bridge and Roads
And that himself his Executors Administrators
and Assigns is and shall be entitled to the Profits and
Advantage of such Share Given under the Common
Seal of the said Company the Twenty fifth
Day of January in the Year of our Lord One
Thousand Eight Hundred & fifteen

Ent.ᵈ David Ensign

Clerk to the Company
of Proprietors.

A 19th-century share certificate *illustrates the undertaking in which*
Isaac Galpin *was purchased* 'one one hundred and fifty fifth share'.

The great virtue of the turnpike system was that it led to the establishment of turnpike trusts and, through them, to the appointment of full-time Surveyors responsible for the maintenance of the roads. This in turn led to a gradual improvement in status and responsibility of the Surveyors and the beginnings of a professional approach to their duties by the salaried officials.

Daniel Defoe, whose main fame rests on his authorship of *Robinson Crusoe*, provides a valuable source of information in his account of his tour through England and Wales in the second decade of the eighteenth century. By this time the turnpike system had become established and was being rapidly extended. The improvements due to the new system were evident although much remained to be done, and many roads remained in a thoroughly bad condition. Defoe does not hesitate to describe these as, for instance, in his description of the roads in Kent in 1722.

'I left Tunbridge . . . and came on to Lewes, through the deepest, dirtiest, but in many ways the richest and most profitable country in all that part of England. The timber I saw here was prodigious . . . sometimes I have seen one tree on a carriage, which they here call a tug, drawn by two and twenty oxen, and even then 'tis carried so little away, and then thrown down, and left for other tugs to take up and carry on, that sometimes 'tis two or three years before it gets to Chatham; for if once the rains come in it stirs no more that year, and sometimes the whole summer is not dry enough to make the roads passable.

'Here I saw a sight which indeed I never saw in any other part of England: Namely, that going to church at a country village, not far from Lewes, I saw an ancient lady, and a lady of very good quality, I assure you, drawn to church in her coach with six oxen; nor was it done in frolic or humour, but meer necessity, the way being so stiff and deep that no horses could go in it.'

Although the need for better roads was evident to everyone, the imposition of the toll led in some parts of the country to riots in which the turnpikes and toll-houses were destroyed. Many

The road through the Pass of Llanberis *in North Wales was winding and deeply rutted.*

such acts of violence occurred in the years about 1750 in such places as far apart as Bristol, Northampton and Leeds, these being put down with the assistance of troops with fighting and bloodshed.

The provision of an organization to provide funds and administer the road system did not itself make better roads; it still remained to work out the technical problems of road construction. The methods adopted by the Romans were not appropriate as

31

they depended for their success on an abundant supply of military or slave labour and although labour in the eighteenth century was still cheap by modern standards, it was much too expensive for building roads of the Roman type. A lead in new methods was given by the French where the central government had taken the matter up as early as the mid-seventeenth century and had in 1720 established a body of civil servants to oversee bridges and roads. The need to train men for this service led to the formation by the mid-eighteenth century of a state-run school—the École des Ponts et Chaussées, which was to have a profound influence on civil engineering during the following century. Early work by the school greatly improved the French roads but a great step was taken by P. M. J. Tresaguet about 1764 when he constructed the first well-engineered road of modern times depending on two essentials which are still recognized—an impervious surface protecting a dry bed. Tresaguet's road consisted of a foundation layer of stones on edge to a height of 6 or 7 inches; on this was a further layer of smaller stones to about the same thickness and the whole was topped with a 3-inch layer of what Tresaguet describes as 'pieces broken to about the size of a small walnut'. This final layer consolidated into a hard, waterproof crust and, with a moderate camber, shed rain-water into the side ditches. Tresaguet claimed that his road surfaces lasted ten years when properly maintained and on this construction the French built a great road system which extended rapidly into most of the European countries.

The construction of roads on the continent of Europe has always been inspired by some degree of military necessity. In Britain, the first real efforts of the eighteenth century to improve

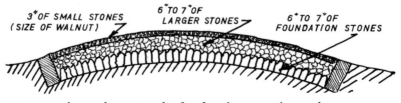

Tresaguet's road consisted *of a foundation with two layers on top, providing the essentials—an impervious surface and a dry bed.*

General Wade's extensive programme *of military road construction in Scotland during the 18th-century campaigns included this bridge at Aberfeldy in Perthshire.*

roads arose from the need for improved communications for the campaigns against the Scots. General Wade, commanding these operations, organized an extensive programme of road construction which was carried out by the troops under the direction of Burt, his engineer officer. Built for purely military purposes, they served for little else, and offered no advantages for social or commercial intercourse for the Scottish people. It was left for Telford, half a century later, to merge them into a comprehensive system of economic value.

One of the first men in Britain to appreciate the requirements of good road construction was Robert Phillips who, in 1737, read a paper to the Royal Society on the state of the highroads in England. He laid down the principle that the road should consist of a self-draining top surface of gravel or similar material on a dry bed, virtually the same specification as that of McAdam later. Phillips' proposals, applied by constructors lacking engineering experience, were not entirely successful and it was a blind man who first applied sound principles to road engineering in England. This was John Metcalfe, of Knaresborough, known as 'Blind Jack', who, although sightless from the age of six years following smallpox, constructed 180 miles of turnpike roads in Yorkshire with conspicuous success. John Metcalfe was described in 1782 by a Mr Bew:

'With the assistance only of a long staff, I have several times met this man traversing the roads, ascending steep and rugged heights, exploring valleys and investigating their several extents, forms, and situations, so as to answer his designs in the best manner. The plans which he makes, and the estimates he prepares, are done in a method peculiar to himself, and of which he cannot well convey the meaning to others. His abilities in this respect are, nevertheless, so great that he finds constant employment. Most of the roads over the Peak in Derbyshire have been altered by his directions, particularly those in the vicinity of Buxton; and he is at this time constructing a new one betwixt Wilmslow and Congleton, to open a communication with the great London road, without being obliged to pass over the mountains. I have met this blind projector while engaged in making his survey. He was alone as usual, and, amongst other conversation, I made some enquiries respecting this new road. It was really astonishing to hear with what accuracy he described its course and the nature of the different soils through which it was conducted. Having mentioned to him a boggy piece of ground it passed through, he observed that that was the only place he had doubts concerning and that he was apprehensive they had, contrary to his directions, been sparing of their materials.'

Metcalfe laid a foundation of large stones, covered it with road material having a cambered surface and drained the surface water into large ditches at both sides of the road. Not only did Metcalfe achieve fame as a road constructor, but, in spite of his blindness, he joined a volunteer regiment and took an active part in the campaign in Scotland in 1745.

It was a blind man who *first applied sound principles to road engineering in England. This was 'Blind Jack' Metcalfe of Knaresborough who, although sightless from the age of six, built 180 miles of turnpike road in Yorkshire in the 18th century.*

Thomas Telford, in his capacity of Surveyor of Public Works for the County of Salop, had much to do with the turnpike trusts of that county. His achievements in that direction, coupled with his rising reputation, led to his being appointed in 1803 to direct the improvement and construction of roads in the Highlands of Scotland in an endeavour to improve the economy of the country. During the next eighteen years, Telford personally supervised the building of 920 miles of new roads in Scotland, together with 1,117 bridges, the whole being done by contract under rigid specifications, the number of contracts being 120. During this period Telford brought the contract system to approximately that which is used today and, although not all the contracts were trouble-free, not one involved the parties in litigation.

The Glasgow to Carlisle road which in 1814 was almost impassable, was reconstructed by Telford and the specification of this road, given as an appendix to his *Life*, is typical of the road construction methods for which he became famous.

Dimensions and Formation of Roadway

The breadth is to be thirty-four feet between the fences; of this, eighteen feet in the middle is to be metalled, and the remaining eight feet on each side are to be covered with gravel. In all embankments, the width at the top is to be thirty feet, and the side slopes to be one and a half horizontal to one perpendicular. In all cuttings above five feet, the width between the lower skirts of the slopes is to be thirty feet; all below that depth to be thirty-four feet. The slopes of all the cuttings to be at the same rates as the embankments. The surface of the road longitudinally, or lengthwise, in all cases where there are cuttings and embankings, are to be formed agreeably to the annexed sections; in cases where the particular sections do not apply, the ascents or descents are in no cases to exceed one in thirty, and the changes from the one to the other to be made in regular curves, to the satisfaction of the inspector. In all cases where there are side-cuttings, and a part of the road on moved ground, the surface of the lower part, or moved ground, is to be higher than that of the upper side, to allow for consolidation, so that the finished road may be of a proper form and level.

In the middle of the road there is a metal bed to be formed in all cases where the ground is nearly level, the metal is to be formed upon the natural surface of the ground, so as to have a curvature of four inches in the middle eighteen feet, and the sides or shouldering to be made with moved ground; but on

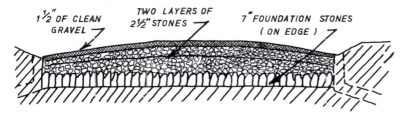

Telford's method of metalling *is revealed in the modern country road (opposite)—two beds are topped by a layer of binding gravel.*

no account is the metal bed to be cut out of the natural ground, unless it is loose gravel or rock. The metalling is to consist of two beds, or layers; that is to say, a bottom course of stones, each seven inches in depth, to be carefully set by the hand from the bottom course downwards, all cross bonded or jointed, and no stone to be more than three inches wide on the top. These stones may be of good whinstone, limstone or hard free-stone; the vacuities between the said stone to be carefully filled up with smaller stones, packed by hand, so as to bring the whole to an even and firm surface.

The top course or bed is to be seven inches in depth, to consist of properly broken stones, none to exceed six ounces in weight, and each to pass through a circular ring two inches and a half in their largest dimensions. These metal stones to be of hard whinstone; the quality of both bottom and top metal to be determined by the inspector. In every hundred yards in length on each side of the road, upon an average, there is to be a small drain from the bottom layer to the outside ditch . . .

Where the height of the embankment shall exceed three feet, they are to stand from one to three months, in proportion as they increase in depth, as shall be determined by the inspector.

Over the upper bed or course of metal there is to be a binding of gravel, of one inch in thickness on an average. In the cross section of the roadway there is to be a curvature of six inches in the middle eighteen feet, and from that on each side, a des clivity, at the rate of half inch to a foot, to within eighteen inche-of the fences. In the remaining space of eighteen inches there is to be a curvature of three inches; making in all about nine inches on each side below the finished roadway.

In passing morassy ground, all the surface upon which the road is to be placed, that is to say, between the fences, is to be brought to a curvature of twelve inches in the middle, and all to be secured with two rows of good swarded turf; the lower one to be laid with the swarded side downwards, and the upper one with it upwards. The metalling upon the said mossy ground is to be made twenty feet in width. On all mossy ground there is to be cut along each side of the road a drain four feet wide at the top, eighteen inches at the bottom, and three feet deep. . . . In all cases the direction of the road is to be set out by the inspector, and all the curves made to his satisfaction; the road is always to be formed for a quarter of a mile, and examined by the inspector previous to any bottoming being set upon it; he is also to be satisfied with the bottoming before any top metal is put on, and with the top metal before any binding is put on.

Besides the sections of the principal cuttings and embankings which are annexed to this contract, there are various small irregularities of from two to three feet, which must be cut down and filled up, in order to bring the road to a uniform surface, no where exceeding one in thirty.

None of the cross drains to be covered until the inspector has examined and satisfied himself with the sufficiency of the pavements and side-walls; and the covers are also to be examined and approved by him previous to the turf being laid on.

In passing through inclosed fields, the contractor is to keep up the cross drains until the road has been completed and taken off his hands, and also to pay the damages to fields or grounds incurred by getting materials for roads or other works, or carrying them by temporary roads.

Telford laid great emphasis on the importance of leaving the natural surface of the ground unimpaired; he preferred the interlocked, slightly elastic medium of a root and soil formation to the alternative of an unknown layer lying below. His foundation, carefully packed with smaller stones, was evidence of his early upbringing as a stonemason.

Telford's greatest road work was the improvement of the Holyhead road. This link with Ireland through North Wales had been a constant source of irritation to the general travelling public and in particular to the Irish members of Parliament, whose complaints eventually led the House to appoint a Commission. As so often the case, this led to nothing; but John Foster, the Chancellor of the Irish Exchequer, obtained Government authority in 1810 to engage Telford to survey the routes between Holyhead and Shrewsbury, also from Bangor to Chester and report on a line for a mail-coach road. This report was completed by April 1811, and submitted to the Committee of the House of Commons. Even then nothing was done until 1815, when Sir Henry Parnell, Member for Queen's County, pressed the Government into forming a Board of Parliamentary Commissioners to administer the construction and improvements on the road from London to Holyhead, Telford being appointed their engineer. The existing road from London to Shrewsbury (153 miles) was treated as a separate section with one assistant acting under Telford. It was still administered by seventeen turnpike trusts which were allowed to remain effective, subject to safeguards ensuring the proper maintenance of the newly made improvements. The Shrewsbury to Holyhead road of 107 miles required much more extensive improvements and was under the supervision of an assistant engineer with four inspectors. The Welsh section of road had previously been managed by seven turnpike trusts, but on Telford's recommendation these were bought out and the administration vested in the Parliamentary Commissioners.

By 1819 the road had been greatly improved and in that year Parliament sanctioned the crowning contribution to Telford's great work, a bridge over the Menai Strait and a new road across Anglesey. The improvement of the road from Bangor to Chester, including a bridge at Conway, was also authorized. These works were completed by the end of 1825, thus providing between London and Holyhead a road which, for alignment and gradient, could not be surpassed today. Now classified as the A5, with no slope greater than one in twenty, it is a top gear road and the motorist using it today should thank Telford, who planned and constructed it nearly a century and a half ago.

Telford was a perfectionist in road engineering, and his Rolls-Royce standards cost more than some road authorities were prepared to pay. For them, came the Henry Ford of road engineers—John Loudon McAdam. Whereas Telford took road engineering as a part of civil engineering, McAdam was a highway engineering specialist. Incidentally, John Rennie considered road building to be an inferior art, and steadfastly refused to undertake any such work as being beneath his dignity.

John Loudon McAdam, born in Ayrshire, went to America and made a fortune, which he brought back to his native county.

The Strand was repaved in 1851. This contemporary print shows the foreman beating time for the ramming of the paving blocks.

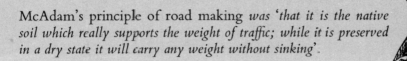

THREE LAYERS OF 2" STONES   IMPERVIOUS

McAdam's principle of road making *was 'that it is the native soil which really supports the weight of traffic; while it is preserved in a dry state it will carry any weight without sinking'*.

There he became a Deputy-Lieutenant and a road trustee, which activity led him, rather belatedly, to his main interest in life. During his Ayrshire days he experimented with road formations and arrived at conclusions which greatly reduced the first cost of road construction. He became Surveyor-General of the Bristol roads in 1815 and in 1827 occupied the same post for the central government. During these years he wrote extensively about his road-making ideas, his books being translated into several languages, thus bringing engineers to Britain from many countries and spreading the influence of the McAdam method.

The fundamental principle of McAdam's roads was expressed in his own words 'that it is the native soil which really supports the weight of traffic; that while it is preserved in a dry state it will carry any weight without sinking'. To maintain the dryness of his foundation, McAdam relied on adequate drainage, the raising of the road above the water-table, an adequately cambered top and, most important of all, an impervious wearing surface. He dispensed with Telford's heavy stone foundation, but relied on the principle of consolidation. His stones were gauged in a 2-inch ring and their individual weight limited to 6 ounces; they were put down in three layers of about 4 inches each, well packed and rammed to the correct camber before the next layer was laid. There was no gravel blinding, as McAdam found that the iron tyres of coaches and wagons ground off chips which soon packed into the voids of his top surface, making the whole completely watertight.

41

The work of Telford and McAdam came at a time when both Europe and America were undergoing rapid expansion. Coach travel was speeding up—the new roads enabled a horse to treble its hauling power. By the mid-nineteenth century the civilized world was covered with a network of metalled roads and the word *macadam* had passed into the language. McAdam's sons continued his work and many toll roads in England were maintained by them after his death in 1836.

Telford's methods of road construction, although more expensive at first, have proved cheapest in the long run and while many modern roads still exist on Telford foundations, it is doubtful if any original McAdam roads can have survived as long. It must be appreciated that although many roads were constructed in general accordance with the principles taught by McAdam, their execution was not always up to his standards and they required constant and costly repair.

The development of Portland cement by Aspdin and Johnson in the first decades of the nineteenth century was bound to lead highway engineers to consider concrete as a road material and so, in the 1850s, the first concrete roads appeared, in Austria first, then in England in 1865, followed by other Continental countries and America. Other materials in the form of tar and asphalt, came into use in the 1830s. Tar and stones, mixed while the tar was hot, were used for the top layer of McAdam roads, the material was rolled after cooling and a layer of sand applied to fill the voids. Known later as tarmac, it was first used in Nottinghamshire in the 1830's, but, like asphalt, its fullest application had to wait until the coming of the rubber-tyred motor-car which sucked out rather than rolled in the fine binding material of the water-impervious macadam roads.

Although asphalt was used as a mortar for building purposes in the Middle East some 5,000 years ago, and even at that time for binding stone slabs in road construction, there is no evidence of

The Highway of the Sun *between Bologna and Florence almost ignores the difficulties of the terrain which are clearly indicated by the course of the old road.*

its use as a wearing surface until the middle of the nineteenth century. The French engineers of roads experimented in the eighteenth century with asphalt as a binder, and this work led de Sassenay to develop its use in the 1830s from local French sources of rock asphalt and bitumen. Telford also had the idea of making a waterproof surface with the addition of an 'admixture of a moderate quantity of inpenetrable materials' which appears to have been the subject of an experiment on the Holyhead road near Wolverhampton after his death. The use of powdered asphalt led eventually to the application of the material in a hot condition as a waterproof surfacing and it was thus introduced into the United States in the 1870s by De Smedt, a Belgian.

The roads of Britain have been described as a palimpsest of history. This description may be applied to the whole civilized world as in each case the development of a country is written in the road pattern of successive decades and centuries. One could mention the 1,500-miles long Grand Trunk Road of India which runs from Calcutta to the North-West Frontier, initially built for the quick passage of marching troops, and tree-lined to afford shade from the burning Indian sun; or the Burma Road of the last war built almost literally by hand to connect North Burma with Chunking; or the great toll roads of America whose multi-carriageways connect states over routes once traversed by the covered wagons of the pioneers. Even a short journey on British roads takes the traveller through time as well as space. Across the crinkled parchment of the countryside, folded by nature and scratched by man, he may, within a few miles traverse roads first laid down by neolithic man, then by the Roman, Danish, Saxon or Norman invaders, roads built for pilgrims, trade or war, diverted for the convenience of landowners or the Enclosure Commissioners, planned by turnpike trusts, by Telford, McAdam or one of the thousands of anonymous contributors to the pattern of our road system. And the lines are still being inscribed in the bold strokes, slowly but inevitably being drawn across the whole countryside and the world by engineers today to provide for the needs of the fast, heavy, power-driven vehicles of our own times.

Early canals in Egypt, Babylon and Europe
the first ship canal in Britain — Leonardo da Vinci
canals in 17th-century Britain — Brindley and
the Bridgewater Canal — the Grand Trunk Canal
Smeaton's Forth–Clyde Canal — Telford's canal
engineering — the aqueducts of Chirk and
Pontcysyllte — the Caledonian Canal and the
Gotha Canal — de Lesseps and Suez — the Kiel Canal
and European trade — the Manchester Ship Canal
the Panama Canal and the St Lawrence Seaway

◀The Corinth Canal *was a costly undertaking as* $2\frac{1}{2}$ *of its 4 miles were cut through solid rock to an average depth of 190 ft. The Canal was a technical success but little used by shipping.*

Rivers and Canals

THE USE OF rivers as means of inland communication is of great antiquity; even today, the most undeveloped races usually possess some skill in the use of canoes or boats, and it is reasonable to assume that similar knowledge had been acquired by man many thousands of years ago. A regular supply of water was necessary for the well-being, not only of man himself, but also of the animals he hunted and those which he ultimately domesticated, and for the plants which supplied him with food or shelter. His need for this water confined his activities to its neighbourhood and it was therefore inevitable that he would eventually use it to assist him in moving himself or his goods from one place to another.

The development of trade added importance to the use of water for communications, while the cultivation of crops and accumulation of herds of beasts called for some means to regulate the supply of water during the dry period of the year. Irrigation canals in Egypt were normal to their way of life as early as the time of Rameses, who succeeded to the throne in 1306 B.C. The ancient historians assert that there were as many as eighty canals in Upper and Lower Egypt, including the Grand Canal between the Nile and the Red Sea, a length of 37 miles, having a width of 100 feet and 40 feet depth—approximately the under-water section of the Cunard liner *Queen Elizabeth*. The Babylonians also used canals for trade and irrigation, in fact, the improvement of rivers and the cutting of canals was a branch of civil engineering knowledge which spread simultaneously with the Mediterranean trade. The Romans included the building of canals and the improvement of rivers in their engineering, the general Drusus connected the Rhine with the Issel, Nero started a ship canal through the isthmus of Corinth but failed to complete it and the same fate overtook the efforts of Lucius Verus in his attempt to join the Moselle and Rhine. In Britain, Agricola cut the Caer Dyke from the Nene at Peterborough to the Witham at Lincoln, extending it by the Fossdyke from Lincoln to the Trent at Torksay, although this 7-mile extension is attributed by some authorities to Henry I, who, in any case, partially restored it to navigation in 1121. It may thus be considered the first British canal. This discounts a canal known as the *Kingsdelf* in Hunting-donshire, referred to in the Anglo-Saxon Chronicle, as this was probably an irrigation ditch. The Romans are also reputed to have cut a canal connecting the Itchen to Winchester, but it is unlikely that they found it necessary to do more than improve the existing river, which no doubt they did in many other rivers of Britain.

Canals were also cut and rivers improved by the early civiliza-tions of the East; there is evidence that the waters of the Ganges were used for irrigation through artificial cuts. In China, the use of water as a means of transport and for agriculture has been extended by canals and irrigation channels from time immemorial and on such a vast scale that in many places a great proportion of the population lives afloat. The Imperial Canal of China con-necting north to south, was completed in the year 1289. For centuries this great waterway, over 1,000 miles long, has con-nected Canton to Pekin 825 miles apart, a means of transporting goods from many parts of the world to the heart of China.

These early canal systems were in every case a part of an advanced civilization, whereas the use of natural rivers never ceased to be an important, and sometimes the only, means of transport. In Britain, as in most European countries, coastal ships carried goods between the main ports and often up to the inland ports, many of which subsequently silted up, sometimes as a result of mineral workings up the very rivers which they served. The founding of Plymouth is a typical example; the old rhyme says: 'Plympton was a Borough Town when Plymouth was a fuzzy down', and not only a borough, but the port serving im-portant tin mines for several miles around. Ultimately, the wash-ings of the tin works, or stannaries, so silted up the estuary of the

As the Plym estuary became silted up, *Plympton was replaced by a new harbour at Sutton, later Plymouth. Rendell's Lary Bridge spanned the river below Plympton.*

river Plym, that a new harbour was founded at the village of Sutton, from which the port of Plymouth grew. So in the same way did other industries such as collieries, china clay, iron and copper works rob themselves of their convenient means of water transport.

The use of rivers for navigation led to a conflict of interests between boat operators and millers. Even in Roman times, water was exploited as a source of power and, by the Norman Conquest, mills were abundant on the rivers of Britain. In most cases, however, a water mill depended on a mill dam across the river which became an obstruction to navigation, and to permit the passage of boats, mill dams on navigable rivers were often provided with removable openings, or *stanches*, usually under the control of the miller. It is not difficult to imagine the battles which arose when, in dry seasons, the millers refused to open the stanches except at intervals of several days, in order to conserve the precious water power for the mills. They were powerfully aided in this as more often than not the mills were held as a perquisite of the Lord of the Manor.

The construction of stanches is described by David Stevenson in his *Canal and River Engineering*. He quotes Sir William Cubitt:

'when he undertook the improvement of the Stour in Essex, there were thirteen stanches along the course of the river. These stanches consisted of two substantial posts, which were fixed in the bed of the river, at a sufficient distance apart to permit a boat to pass easily between them, and connected at the bottom by a cross cill. Upon one of those posts was a beam turning on a hinge or joint, and long enough to span the opening. When the "stanch" was used, the boatmen turned the beam (which was above the level of the water) across the opening, and placed vertically in the stream a number of narrow planks resting against the bottom cill and the swinging beam, thus forming a weir which raised the water in the stream about 5 feet high. The boards were then rapidly withdrawn, the swinging beam was turned back, and all the boats which had been collected above were carried by the flow of water over the shallow below. By repeating this operation at given intervals, the boats were enabled to proceed a distance of about 23 miles in two or three days.'

The battle between navigation and mill interests cannot be better illustrated than by the early struggle for rights of water access from the sea to the City of Exeter, which led ultimately to the establishment of the Exeter Canal, the first ship canal in Britain and the first to be provided with pound locks. Previous to the reign of Henry VI, the tide came up as far as Exeter, so that barges and small craft could navigate up the Exe to the watergate of that city. About that time, Isabella de Fortibus, Countess of Devon, built a weir, called after her 'Countess Weir', still known by the same name and passed in the summer by thousands of motorists every day on the Exeter by-pass road. In this weir an opening of 30 feet was left for the passage of vessels, but in the twelfth year of Edward I, this passage was blocked.

'Trew's Weir' *was the result of the struggle for rights of water access from the sea to Exeter and led in its turn to the establishment of the Exeter Canal.*

In addition, other weirs were added by the Courtenays, Earls of Devon. Legal proceedings were taken by the citizens of Exeter, who gained verdicts against the Courtenays, but the power of the earls was greater than that of the law. In the thirty-first year of Henry VIII (1540) an Act of Parliament was obtained for the restoration of the navigation, but although many efforts were made and much money spent in that and the two succeeding reigns, the citizens did not succeed in bringing the river back to its original state.

In 1563, the city engaged 'John Trew, a Gentleman, of Glamorganshire, in Wales', as their engineer. Trew, instead of clearing the river, made the city accessible to it by a canal having a lock. This was a true pound-lock canal, similar in all essential points to canals of the present day. Trew originally proposed to form a canal by placing a lock at the lower end of an existing mill-leat on the east side of the river, which in view of the high banks, was probably the only possible route on that side. He finally decided, however, to dig a new canal on the west side. He added an additional weir above those already in existence which is named after him 'Trew's Weir' or, alternatively, 'St Leonard's Weir'.

In 1699, William Bayly undertook the 'widening and digging the new canal or river, making a stone wear, and digging the broad from the key thereunto'. Bayly did enough work to make the canal impassable and then ran away with the money. The city then applied for an Act of Parliament and accompanied their application with a map, plans and an estimate of £13,340. The opposition of petty interests in Exeter and Devon succeeded in getting the Bill rejected but the city fathers persevered for twenty-five years with their attempts to obtain Parliamentary sanction. In the meantime they managed without an Act to restore the canal to a depth of 10 feet and extended it to Topsham, a work which was completed in 1703. In 1819, James Green, an Exeter engineer, who had done excellent work on other canals in the West of England and South Wales, was called in by his city authorities to submit proposals for improvements to the canal. He reported that the canal was in a very imperfect condition, the banks being very much 'washed down'. He recommended a great increase in the cross section of the canal, a depth of 15 feet, new locks, and an extension to an additional lock at Turf. The excavation for this new lock gave trouble through water pressure at high tide, but by a system of culverts to relieve the pressure, the work was completed successfully. Telford saw the work in progress and said that 'he had never seen so troublesome a foundation, and he highly approved of the method adopted for preventing the upward pressure of the sub-water'. Green's improvements to the Exeter Canal have been maintained to the present day, enabling Exeter to handle small ships and to be considered as a seaport.

The invention of the system of chamber and gates which we know as the pound lock, or more commonly, the lock, has been variously attributed to the Italians and the Dutch. It has been argued that the Italian claim is strengthened by the more hilly nature of the country, making changes of level necessary at more frequent intervals, but against this argument it must be remembered that most early Dutch canals were dug in land which was below sea level and therefore locks would be required to gain access to the sea. According to Cresy, locks were invented by Dionisius and Pietro Domenico of Viterbo in 1481. A lock is reputed to have been built in 1488 on the River Brenta, near Padua, and locks were used by Leonardo da Vinci in 1497, who supervised the building of six uniting the two canals of Milan. The change of level here was 34 feet. One of these canals was the Naviglio Grande, which was constructed between A.D. 1197 and 1257 from Milan to the Ticino River, a distance of 31 miles. On this canal the water level was maintained by dams and the transfer of barges from one level to another was achieved by pulling them over inclined planes by winches. The original pound lock consisted of a lock chamber of brickwork or masonry, closed at the upper end by a pair of mitre gates, and at the lower end by a single gate, hinged on one side.

A lock is reputed to have been built in 1488 on the River Brenta, near Padua, and in 1497 Leonardo da Vinci supervised the building of six uniting the two canals of Milan.

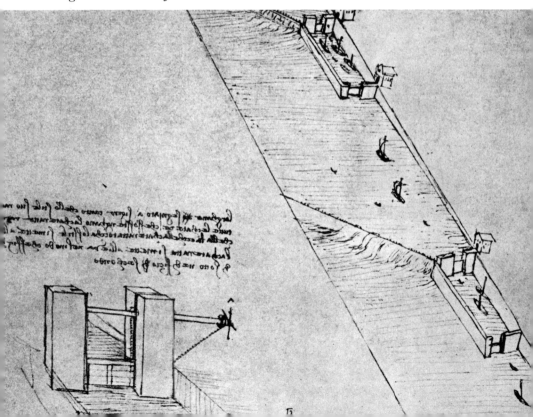

The original pound lock *was closed at the upper end by a pair of mitre gates and at the lower end by a single gate (detail above), hinged on one side. The engraving opposite, from Zonca's 'Nuovo Theatro di Machine, et Edificii' (1597) shows the mitre gates at the lower end.*

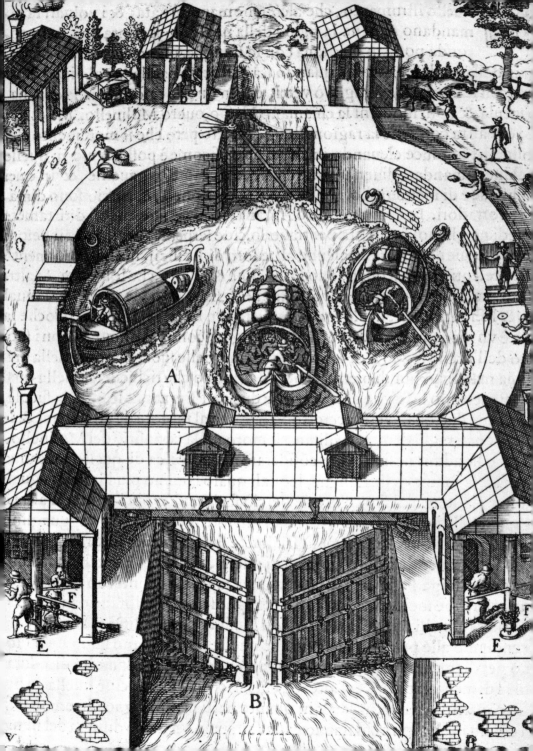

Leonardo's work on the Milan canal project is described in a biography published with his *Treatise on Painting*, the English translation being dated 1721.

'About this time', says the biographer, 'Duke Lewis formed a design of supplying the City of Milan with water by a new canal. The execution of this project was deputed to Leonardo, and he acquitted himself of the trust in a manner that surpassed all expectation. The canal goes by the name Mortesana, being extended in length above 200 miles: and, navigable throughout, it passes through the Valteline and the valley of Chiavenna, conducting the waters of the River Adda to the very walls of Milan, and enriching both the city and the adjacent Campaign by its communication with the Po and the sea. This was a noble and a difficult enterprise, every way worthy of Leonardo's genius. He had here several difficulties to grapple with in digging the ancient canal which conveys the waters of the Tesino to the other side of the city, and which had been made above 200 years before, while Milan was a republick. But Leonardo surmounted all opposition, and happily achieved what some may think miraculous, rendering *hills* and *valleys* navigable with security.'

In the Low Countries and France, there was a great canal building activity in the sixteenth and seventeenth centuries. The Brussels canal, connecting that city to the Scheldt, was completed in 1560 and the Burgundy and Picardy canals were cut in the first half of the seventeenth century. The greatest French work of that century was, however, the great Languedoc Canal, connecting the Atlantic with the Mediterranean; commenced in 1666, it was completed in 1681. It is 180 miles long, 144 feet wide, and 6 feet deep. It was built by Riquet and has locks 102 feet long and 19 feet 8 inches wide.

Although in Britain canal building lagged behind the Continental countries, the seventeenth century brought about a great increase of interest in the improvement of rivers. This activity had already started in Tudor times and Acts had been passed for improving the Severn, Thames, Tyne, Yorkshire Ouse, Humber,

Lea and other rivers. These had led to considerable improvements in the rivers, but on 16 July 1618, a Patent was granted to Captain John Gilbert for the first dredger, described as 'a water plough for the taking upp of sand, gravele, shelves, or banckes out of the river Thames and other banckes, harbours, rivers, or waters'. This invention seems to have been put into use, as Captain Gilbert, in conjunction with James Feese, a London merchant, obtained a further licence for the machine's manufacture in July 1629.

The improvement of rivers was further extended by the activities of Andrew Yarranton who, from being a linen-draper's apprentice, successively became a captain in the parliamentary army, a farmer and the owner of an iron works. During that chequered career he had travelled in Holland and France and in 1677 published a treatise *England's Improvement by Sea and Land, to Outdo the Dutch without Fighting*. This work, completed in 1698 by the publication of a second part, showed the advantages to be gained in Britain by following the example of France and the Low Countries in developing water communications. Yarranton's writings came at a time when the country's economy was expanding, the towns were growing, and transport had become a serious problem; he was therefore consulted on matters of river improvement and eventually became the equivalent of a modern consulting engineer. As such, he surveyed a number of rivers for large landowners and for them supervised the execution of improvement works which were so successful that, over a period of sixty-four years, Acts were passed for improving sixteen rivers, which were partially navigable, and for opening up ten more which had not previously been usable for navigation. Not all these Acts became effective, but the influence of Andrew Yarranton on inland navigation was a major factor leading up to the canal expansion of the eighteenth century.

In 1755, as a result of activities attributed to the 'commercial and enterprising inhabitants of Liverpool', an Act was passed for the improvement of the Sankey Brook, from the Mersey to St Helens. Instead of improving the river, the projectors cut a canal with locks, the first in Britain since that of John Trew in 1563. Before the Sankey Brook Canal was completed in 1760, other, and greater, activities had started, which were to lead ultimately to the great canal expansion in Britain which took place in the next half-century or so. At the time of their activities with the Sankey Brook, the energetic citizens of Liverpool were investigating the possibility of connecting the rivers Trent and Mersey with a canal. Two surveys were made, and on one of these was engaged a self-taught millwright, James Brindley. Born in 1716, near Buxton, Brindley apprenticed himself in 1733 to Abraham Bennett, a master millwright for a term of seven years. Although he was virtually illiterate, a handicap which he only partially overcame, Brindley soon proved himself to be a natural-born engineer and, in a few years, had gained for himself a reputation for the solution of mechanical problems. It was this reputation which led to his introduction to the Duke of Bridgewater, who was to be Brindley's sponsor in the first of a series of canal projects which were to change for the better the transport system and economy of industrial England.

Francis Egerton, the third and last Duke of Bridgewater, was born in 1736 and succeeded to the title at twelve years of age through the death of his older, consumptive brothers, although he, a sickly child, had been neglected by his nurses, perhaps to his advantage. At seventeen, he embarked on the 'grand tour', from which he returned restored to health and with, no doubt, some first-hand knowledge of Continental waterways. Much of the Duke's property lay in the rich industrial districts of Lancashire and his estate at Worsley covered extensive coal measures. In these matters the young Duke at first took little interest until an unsuccessful love affair from which he emerged a sadder but a wiser man. From about 1759 onwards, he devoted himself to the improvement of his estates and, in particular, to the development of his mineral rights.

At that time, the textile industry was advancing rapidly in Manchester which, from an industry of home workers, was developing into a town of factories. With the increase of population arising from this development came problems of housing, feeding and warming great numbers of people. The road system of that part of England was bad, pack horses being used exten-

The first canalized river *of the Canal Age in Britain was Sankey Brook, from the Mersey to St Helens. Across it the Sankey Viaduct carried George Stephenson's Liverpool and Manchester Railway.*

sively for the carriage of goods into and out of Manchester. The Irwell and Mersey rivers were in the hands of monopoly owners who used their powers to fleece traders of the highest dues possible; even then, in time of drought or flood, the rivers were unusable for many weeks at a time. Such conditions led to great hardship in Manchester and, eventually, to riot; not only was it difficult to bring in food and fuel, but the raw material and finished products of trade moved only with great inconvenience and expense.

Thus the local conditions favoured the Duke when he resolved to cut a canal from his Worsley colliery into Manchester and for this purpose, in 1759, he promoted a Bill which received the Royal assent in March of that year. It is probable that, up to this time, no engineer had been employed in connection with the project as, in those days, Parliamentary Bills and Letters Patent could be obtained for proposals of a vague nature and without necessarily being supported by surveys or drawings. The Duke

The 3rd Duke of Bridgewater (*left*) invited James Brindley to build a canal from his Worsley colliery to Manchester. The Barton Aqueduct seen in both of these portraits was to enable the canal to be carried on the level without the need for locks.

therefore on the advice of his land-agent, John Gilbert, invited James Brindley to undertake the survey, lay out the works and supervise their construction. Brindley was known to Gilbert through his brother, Thomas Gilbert, agent to Lord Gower, on whose behalf Brindley had been consulted in connection with a canal to connect the rivers Mersey and Trent.

Brindley made what he described in his notebook as 'an ochilor servey or a recconitoring' and made recommendations to the Duke from which it was clear that the powers obtained by the 1759 Act would be insufficient. Brindley's proposal was to carry the canal, as far as possible, on the level. This meant crossing the Irwell not by locking down to river level and then up on the other side but by constructing a substantial embankment across the low ground on the north side of the river and crossing the river by a large aqueduct of stone, a scheme so apparently absurd that public opinion considered the Duke and Brindley insane to pursue it. The Act, however, was passed in 1760 and construction proceeded. Not only the embankment and a tunnel into the mine at Worsley but also the aqueduct were entirely successful, and on 17 July 1761 the canal was opened by the first boat passing along its whole length. The opening was followed by an extension to Runcorn which provided greatly improved navigation to Liverpool.

These canal works, unimportant as they may appear against works since undertaken, were, in fact, a turning-point in civil engineering, including as they did so many technological advances which were later exploited on a much greater scale, not only by Brindley himself, but by other canal engineers, and, still later, by the generations of civil engineers who succeeded them. The success of the canal also stimulated the Duke himself to further efforts and, for the remainder of his life, he interested himself in mines, canals and mills. He lived until 1803, surviving Brindley by thirty-one years, and in that time seeing the canal system of Britain grow beyond even his most extravagant conceptions. He remained unmarried and, at times, had every penny of his resources locked up in his projects. His canals alone absorbed £200,000 of his money. They returned a handsome profit, however, ultimately bringing in a revenue of £80,000 per annum.

The successful Bridgewater Canal aroused great public interest, not all of it favourable. The opponents of canals said that they would ruin the trade of those earning their livelihood on the roads of Britain and those engaged in the coasting trade, that they would seriously reduce the number of draught horses and, by cutting great areas of land into waterways, would destroy areas valuable for corn growing. The Navy would suffer from the reduction in coastwise shipping, and the river navigation would be neglected as a result of the artificially cut waterways. These objections inspired Richard Whitworth, a contemporary of the Duke, to reply, which he did in a treatise published in 1766 under the title *The Advantages of Inland Navigation*. In this essay, Whitworth proposed a system of canals to link 'the three great ports of Bristol, Liverpool and Hull'. He points out that rivers 'are subject to floods at one part of the year, and at the other to shallows for want of water in a dry summer', maintaining that 'that sort of navigation is almost universally agreed to be laid aside'. In stressing the military advantages of canals, he cites the difficulties experienced in 'the inconsiderate rebellion of 1745' owing to the foul condition of the highroads, 'even four score miles from the metropolis'. In an attempt to meet the objections of the established road transport operators, he makes the very short-sighted suggestion that they be protected by Act of Parliament prohibiting any canal being made nearer than 4 miles to any town or village. Whitworth proposed a canal running from Tern Bridge near the Severn in Shropshire, through Bridgeford, Stafford and Burton, to join the Trent at Wilden Ferry, with another arm running from Bridgeford through Madeley Park in Staffordshire to join the Weaver at Winsford Bridge in Cheshire, thus connecting with the Mersey.

Although Whitworth's scheme was never adopted, the ports of Liverpool, Hull, and Bristol were eventually connected by the Grand Trunk Canal of James Brindley, of which the first sod was cut on 26 July 1766. Brindley had already been engaged on a survey of the Trent and Mersey section of this canal before his introduction to the Duke of Bridgewater, and, following his successful completion of the Duke's canal, was engaged to collaborate with John Smeaton in making a joint survey and report. Great difficulties were put in their way by opponents to the scheme who included objectors to canals generally and, also, the more powerful promoters of rival schemes, who resented the idea of the Duke of Bridgewater obtaining a monopoly of canals in the important industrial areas to be served. The Bill became law, as it was strongly supported by other interests, one of the most active being Josiah Wedgwood, who cut the first sod at the inaugural ceremony on 26 July 1766. Wedgwood moved his whole works from Burslem to Shelton, on the bank of the Trent and Mersey section, where he built the finest factory then known in Britain, which he named Etruria. He built cottages for his workpeople and, characteristic of the ideas of his times, a

mansion for himself on the higher ground immediately over-looking the factory. The great factory-owners of those days did not waste time in commuting. Whitworth's idea of connecting the Mersey, Humber and Severn was brought to fruition by the construction of the Wolverhampton Canal, later named the Stafford and Worcester, from the Severn at Stourport to the Trent at Great Heywood. Brindley intended at first to bring his canal to the Severn at Bewdley, but the inhabitants of that town opposed the idea; he therefore carried it to a place called Little Mitton, which, at that time, had only one small alehouse, named Stourmouth, and where he met no opposition. At this place he built his locks down to the Severn, together with a barge basin and warehouses, around which grew a town, Stourport, as a consequence of the canal traffic. This town remains today a picturesque reminder of James Brindley, with its peaceful setting of eighteenth-century buildings around the canal basin, now a central depot for British Waterways.

Samuel Smiles refers to the Grand Trunk Canal as

'the most formidable undertaking that had yet been attempted in England. Its whole length, including the junctions with the Birmingham Canal and the river Severn, was 139½ miles. In conformity with Brindley's practice he laid out as much of the navigation as possible on a level, concentrating the locks in this case at the summit near Harecastle, from which point the waters fell in both directions, north and south. Brindley's liking for long flat reaches of dead water made him keep clear of rivers as much as possible.'

His long reaches of lock-free canal were achieved by following the contours of the country at the expense of distance. It may be that his lack of experience in lock construction influenced him in this, but it is also probable that he was greatly concerned with the necessity of conserving water, so that his canals would remain effective in times of drought, and this avoided the losses inherent in any system of locks. Brindley was not afraid to use tunnels and the Grand Trunk had five; the Harecastle of 2,880 yards, the Hermitage 130 yards, the Barnton 560 yards, the Saltenford 350 yards and that at Preston-on-the-Hill 1,241 yards in length. These tunnels were of small section and the boats were propelled through them by a procedure known as 'legging', often done by men employed for the purpose, known as 'leggers', who lay on their backs on the barge deck, or on planks provided for the purpose and thrust the barge forward by a walking movement against the sides or roof of the tunnel. A second tunnel was constructed by Telford in 1827 at Harecastle with a towpath which did away with the need for the slow and arduous task of legging in that tunnel. The cutting of Harecastle tunnel took eleven years, and was not completed until after Brindley's death.

His appreciation of the need to conserve water, and his courage in driving canals through all kinds of country, with soils good, bad and indifferent, led Brindley to take a great interest in the problem of making his canals watertight. He achieved this by developing to a fine art the use of puddled clay. Unworked clay, excavated and placed in a canal bed, is by no means a watertight material; if, however, the clay is mixed with a proportion of sand, wetted and kneaded, it becomes a completely leakproof lining and remains so as long as it is kept in the wetted condition. The kneading of the plastic clay was performed either by labourers who 'heeled' it with their boots, or by animals driven over the surface until the right quality of puddle was achieved. This method of making water-containing structures leakproof is still adopted for many applications and the traditional methods of puddling have not so far been improved on mechanically.

Puddled clay was used by Brindley *to make his canal watertight. The traditional method of 'heeling' the mixture of wet clay and sand is still used today in the core of this modern reservoir.*

Before his death in 1772, at the age of fifty-six, Brindley had planned and laid out a number of other canals including the Coventry and Oxford Canals, which completed the connection of the four great rivers Mersey, Humber, Severn and Thames; the Birmingham Canal, which opened up the industrial midlands to navigation; the Droitwich Canal, linking the salt industry with the system; and the Chesterfield, which opened up the coal, lime, and lead resources of Derbyshire to the Trent at Stockwith. He was also consulted about the Leeds and Liverpool Canal, the Aire and Calder Navigation, the Forth–Clyde Canal, the Southampton–Salisbury Canal, the Lancaster Canal, and the improvement of the Thames Navigation to Reading. Can it be surprising

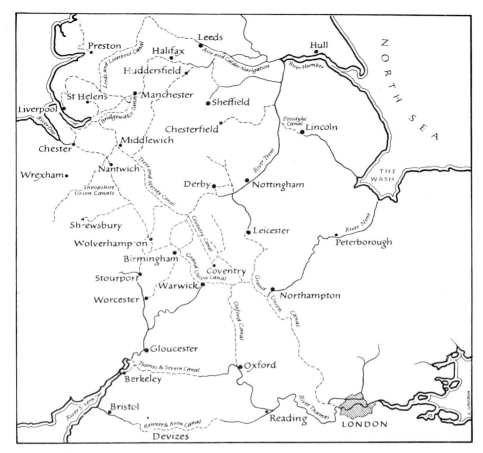

that he died at such an early age, worked out, as were his successors, Brunel, Robert Stephenson, and Locke? His influence on civil engineering in Britain was immense, and his works greatly accelerated the pace of the industrial advance of the eighteenth century. It is safe to say that there is no civil engineer of today who does not derive some of his skill and knowledge from James Brindley.

In 1768, when the first stages of the Grand Trunk system were authorized by Parliament, another proposal was under way in Scotland. For many years schemes had been discussed for short-circuiting the long sea route around Scotland and ultimately various interests combined, not without previous and sometimes

67

acrimonious argument. They invited John Smeaton to survey the various alternatives and submit proposals and estimates for a canal from the Forth to the Clyde. Smeaton's final report and estimates, dated 1767, outlined the route and gave estimates for a canal of 7, 8 or 10 feet depth respectively. The sponsors decided on a canal of 7-foot depth, and it is of interest to compare the estimate with the final cost.

The canal was begun, with Smeaton as engineer, in 1768, but owing to difficulties in raising the money construction was stopped and not until after the issue of a Government loan of £50,000 was it possible to complete it, with Robert Whitworth, who had been a pupil of Brindley, as engineer. It had cost £150,000, very close to Smeaton's £147,337, a very unusual circumstance when most engineers' estimates were, to say the least of it, on the optimistic side. The Forth–Clyde canal was a great financial success throughout the canal era, its receipts rising steadily from £8,000 to £50,000 a year. It was bought by the Caledonian Railway in 1867 for the sake of its Grangemouth Harbour and from that time its traffic declined. It was closed in 1962.

The Forth–Clyde canal was the waterway on which, in 1789, Symington tested his steam propelled boat which was followed in 1802 by the famous steam paddle tug *Charlotte Dundas*. This craft towed two laden barges of 70 tons each a distance of 19½ miles in six hours against a strong wind.

Smeaton was also consulted in connection with other canal works, such as the Grand Canal in Ireland and the improvement of the Birmingham Canal, but his main engineering interests lay in other directions. In one of these, he was akin to Brindley as both men were great experts on millwork, Brindley depending on his flair for the rightness of things based on early experience, but Smeaton, with his better education, made a study of mills which eventually led to the publication of his treatise on mills, the greatest work of its kind.

The great advantage offered to traders by the canals of James Brindley led to their immediate financial success, and money became freely available for the promoters of other canal building

'Charlotte Dundas', *the famous steam paddle tug, in 1802 on the Forth-Clyde Canal towed 2 laden barges a distance of* $19\frac{1}{2}$ *miles in 6 hours against a strong wind.*

projects. The scarcity of experienced engineers led to the development of a new system of organization in which the more eminent engineers acted as advisers or consultants to the promoters, the actual construction and later maintenance being supervised by local men or by assistants appointed to the principal engineer. It was inevitable that such a system led at times to abuses, but, on the whole, things worked out very successfully and the benefits far exceeded the disadvantages. The leading engineers of the time were able to apply their growing knowledge and experience to the greatest possible number of works, while, at the same time, a large number of young men were able, under their general supervision, to gain valuable experience and fit themselves to become worthy members of the rapidly growing profession of civil engineering. Brindley, uneducated as he was, knew little, except what he learnt from the Duke, of the great canal works on the continent of Europe. His successors, however, not only had his experience to go by, but were able to travel and gain knowledge from the great work done by the French and Dutch canal builders.

The later canal engineers found themselves better off financially and were able to plan on more generous lines than Brindley, whose early canal works were conditioned by the Duke's financial limitations and were therefore planned with a view to the greatest economy without the sacrifice of functional effectiveness. This, without in any way reducing the stature of any of Brindley's successors, should ensure him, unlettered though he was, of the highest place in the hierarchy of civil engineers.

The great profits made out of these early canals, and the prosperity which they brought to the towns served by them, led to a great wave of speculative canal promotion which began in 1789, reached its peak in 1792-3, and had faded away by 1797. In the two peak years, thirty schemes for new canals were promoted, some of them sound propositions, but others of little worth, such as the Southampton–Redbridge section of the Southampton–Salisbury canal.

This canal ran into trouble throughout its construction, the length between Southampton and Redbridge, although unnecessary, included a tunnel, which proved so expensive that the shareholders called in John Rennie for a report. The length over the high ground between the rivers Test and Avon was sandy, and the engineer failed to make this watertight—the Brindley recipe for clay puddle was evidently not used to effect. After spending £10,000, the canal was abandoned by the proprietors and only a few traces remain.

Among the advantages offered to engineers by the freer availability of capital was that of constructing their canals along more direct routes. The cost of locks, embankments, aqueducts, and reservoirs was willingly accepted by promoters if it could be shown that time could be saved in delivery. Both Rennie and Telford took full advantage of this. Rennie, whose canals included the Kennet and Avon, the Rochdale, and the Lancaster, achieved some of his finest works of art in the Lune Aqueduct on the Lancaster Canal and his Limpley Stoke Aqueduct on the Kennet and Avon. His great feats of canal engineering include the great tier of twenty-nine locks at Devizes on the Kennet and Avon.

Rennie's feats of canal engineering *include the great tier of 29 locks at Devizes on the Kennet and Avon Canal.*

The Rolle Aqueduct on *the Torridge Canal, near Torrington in Devonshire, blends perfectly with the landscape.*

Benjamin Outram *was the engineer for the Marples Aqueduct on the Peak Forest Canal (left) and probably also the designer of the 'roving bridge' at the junction of the Peak Forest and Macclesfield canals.*

Telford's activities included a great amount of canal engineering and his aqueduct of Pontcysyllte is perhaps the greatest of all the works executed by British canal engineers. The aqueduct is a part of the Shropshire Union Canal system which began as the Chester and Ellesmere canals. The Chester Canal was authorized in 1772 to run from the river Dee at Chester to Nantwich, with a branch to Middlewich, at an estimated cost of £62,000. The canal was a failure; over £100,000 was spent on the main route and the Middlewich branch was not even begun owing to opposition by the Trent and Mersey Canal Company. In 1793, the Ellesmere Canal was authorized by Parliament to cover a route joining Shrewsbury, Chester, Wrexham, Ellesmere and Netherpool, thus joining the rivers Severn, Dee and Mersey. It was intended to have several branches and to serve the whole of the North Wales industrial area around Wrexham. Thomas Telford was appointed engineer.

Work was begun with the Wirral section, built to take barges from the Dee and Mersey and from the Bridgewater Canal and which was opened in 1795 to begin as a successful commercial undertaking. At the same time the construction was begun of the most difficult part of the system, from Chirk to Weston, and the branch to Llanymynech which included the expensive aqueducts at Chirk and Pontcysyllte. Difficulties were met in the country between Pontcysyllte and Chester owing to inaccurate surveys made before Telford's appointment and, in addition, coal was being brought to Chester from new sources of supply. The main canal route was changed to join the previously defunct Chester Canal near Nantwich with the Ellesmere near Whitchurch, leaving the two great aqueducts on what had become a branch. The work was completed in 1805 and the canal opened with ceremonial on 26 November of that year. The full potential of the system was not achieved until 1827, when an Act authorized a junction near Middlewich with the Trent and Mersey Canal. This, with the opening of the Birmingham and Liverpool Junction Canal, authorized in 1826, converted the old Chester Canal from one of minor importance to the best main line canal between Birmingham and the Mersey.

To save the expense *of carrying a puddled-clay channel across the Pontcysyllte Aqueduct, Telford used a cast-iron trough. The 19 arches are also of cast iron, carried on 121-ft masonry piers.*

The two aqueducts of Chirk and Pontcysyllte were designed by Telford to perform the same function as Brindley's original Barton Aqueduct—to save locking down and up the sides of a steep river valley. To save the excessive cost of carrying a puddled clay channel across a high aqueduct, Telford decided to use cast iron in his channels. The Chirk Aqueduct, which consists of ten arches of 40 feet span, carries the canal water surface 70 feet above the level of the river Ceriog and is constructed entirely of masonry with the exception of the canal bed, which is formed of cast-iron plates. The aqueduct at Pontcysyllte is approached on the south side by an embankment 1,500 feet long and 97 feet high where it meets the aqueduct, which consists of nineteen arches of cast iron carried on masonry piers 121 feet high. These arches support the trough, which is constructed entirely of cast iron, made for Telford by his friend William Hazledine whose work in his own craft was fully up to Telford's exacting standards.

Telford was also engineer to the Shrewsbury, the Glasgow and Ardrossan, the Birmingham and Liverpool Junction and the Macclesfield canals, and was responsible for improving the Grand Trunk and the Birmingham canals. His greatest canal works were, however, his ship canals—the Caledonian in Scotland, the Berkeley–Gloucester in England, and the Gotha Canal in Sweden. The difficulty of navigating around the north coast of Scotland had, especially after the 1745 rebellion, led to suggestions for canals connecting the east and west coasts of that country. In 1784, a book published by John Knox advocated the building of canals from Fort William to Inverness, from Loch Fyne to the Atlantic, and from the Forth to the Clyde. His suggestions were based on work already done either as surveys or, in the case of the Forth–Clyde, by the commencement of the work under John Smeaton. The first two are still in use. In 1793, an Act was passed for the formation of the Crinan Canal, which runs from Lochgilphead in Argyllshire to the Sound of Jura, shortening a long sea journey round the Mull of Kintyre to a mere 9 miles. John Rennie was the engineer and it was his first canal. It has a surface width of 66 feet and is 13 feet deep and was completed in 1801 at a cost of £100,000; in 1848 it became a part of the Caledonian Canal undertaking. In 1802, at their request, Telford reported to the Lords Commissioners of His Majesty's Treasury on the cost and practicability of a canal from Fort William to Inverness. This report was based on a survey of the route which he made in the company of William Jessop as consulting engineer and who, over the years, had been Telford's friend and counsellor. It was referred to a Select Committee of the House of Commons with the result

The Caledonian Canal *was designed to overcome the difficulty of navigating around the north coast of Scotland. The coming of the railways removed the original need but it still provides a useful short cut for coastal shipping.*

that in 1803 an Act authorized the construction of the Caledonian Canal and the setting up of a body of Commissioners to administer it. In the same year, and under the powers of another Act, a second Board was formed to supervise the construction of harbours, bridges and roads in the Highlands of Scotland and Telford was appointed engineer to both of these bodies.

In his report of 1802, Telford estimated the cost of the canal at £350,000, but on the passing of the Act, he increased his estimate to £474,000. The design allowed for a depth of 20 feet, as it was intended that the canal should take large merchantmen and warships up to the size of a 32-gun frigate. The route ran through the Great Glen, connecting Lochs Lochy, Oich and Ness; twenty-nine locks were needed to carry the canal over the summit and docks were to be constructed at both ends, one at Clachnaharry being 32 acres in extent.

Many difficulties were met and overcome during the construction of the canal, of which, the ground conditions at Clachnaharry afford a typical example. At this end of the canal, the shore consisted of a very flat mudland into which it was possible to push an iron rod 55 feet. There was also a layer having the consistency of peat, as Telford found it impossible to drive piles for a coffer-dam, the piles rebounding after every stroke of the hammer. He therefore adopted a novel expedient of carrying clay embankments out from the shore into a 20-foot depth of water at neap tides and, at the site of his proposed lock, he dumped a mass of clay over an area exceeding that of the lock. On this clay mass, he placed, as ballast, the stone which was subsequently to be used for the lock construction, and left the whole area in this condition undisturbed for six months. By that time, the mass had sunk 11 feet, and the stone being removed, the lock pit was excavated by digging the compressed mud to a further depth of 8 feet before building the masonry lock bottom and walls. Telford considered that this method of construction worked out to be less costly than a coffer-dam, even if the latter had been possible under normal conditions. The lock at the Corpach end of the canal was founded on rock, and was built inside a coffer-dam constructed within an embankment tipped from the shore to about 100 yards from high water mark and faced with rubble stone.

Although Telford had increased his original estimate from £350,000, to £474,000, the canal cost more than double by the time it was finished, and even then, the depth had been considerably reduced, being only 12 feet at the cuts and 15 feet at the locks. Most of the extra cost was due to the inflation arising from the Napoleonic War, but, in addition to war conditions,

Telford had to meet serious difficulties of supply both in materials and labour as a result of the remoteness of the site. This was probably the reason why he accepted very much lower standards of workmanship than those which he normally demanded. It is also probable that, with his heavy commitments elsewhere, on the Highland roads and bridges for instance, he did not give the Caledonian Canal the amount of oversight which it was usually his practice to provide. The canal was opened in 1822 in its unfinished state and, although it was deepened and reconstructed at a later date, the coming of railways removed the original need of the canal for heavy traffic. It does, however, provide a useful short cut for coastwise traffic and, in the event, probably justified its construction.

Telford was also connected, at one time in its history, with the Gloucester and Berkeley Canal, a ship canal authorized by an Act of 1793 to be constructed between Berkeley Pill on the Severn to Gloucester in order that shipping might avoid the most difficult and dangerous reaches of the river. The story of this canal is one of enterprise and persistence on the part of the proprietors in the face of great difficulties, not of construction, but arising out of bad administration and corruption on the part of some of their servants. The first engineer appointed was Robert Mylne who was to receive a yearly retainer of £350, plus travelling expenses, and to be represented on the site by a resident engineer. The resident engineer appointed was Denris Edson, at 200 guineas per annum, but after nine months he was dismissed to be succeeded by one Dadford, who was engaged for one year only. Troubles developed and the managing committee expressed the view to Mylne that he was not earning his fees and obtained from him an agreement to accept a daily rate of four guineas plus travelling expenses for the time actually spent on the project. The cutting commenced at the Gloucester end with the construction of the dock and continued towards Berkeley. By 1797, however, the money had run out when only $5\frac{1}{2}$ miles of the projected $17\frac{3}{4}$ had been cut, and for twenty years the committee barely kept the project alive. In 1817, however, under the Poor Employment Act, the Exchequer Bill Loan Commissioners, on

Telford's recommendation, provided a loan to enable the work to proceed on the basis of a shorter length of canal terminating at Sharpness. Further difficulties followed as a result of maladministration and another Government loan was made on terms which included a great degree of Treasury control. Under this new administration the canal was at last finished in 1827. It was, at the time, the greatest ship canal in Britain, being 70 feet broad at the water level and 18 feet deep. The original estimate of Robert Mylne was £121,329 10s. 4d. but the total cost worked out at £444,000. The canal greatly improved maritime access to the port of Gloucester and, by the mid-nineteenth century, the docks had become a centre of trade which attracted the attention of the railways. The Birmingham and Gloucester line was extended to the docks in 1862 and, in 1879, the Severn railway bridge was opened to carry Welsh and Forest of Dean coal across to Sharpness, the Gloucester and Berkeley being substantial shareholders in the bridge company. Although the canal never became a great dividend earner of itself, it is safe to say that its existence became and has remained a substantial factor in the prosperity of Gloucester. Even today, the frequent passage of craft carrying oil products is an indication of the value of this link with the work of Mylne and Telford.

Telford's canal activities were not confined entirely to Britain. In 1808, at the instigation of the King of Sweden, he received an invitation from the Count Baltazar von Platen to visit Sweden and report on the possibility of a canal to connect the Baltic with the North Sea. He accepted the invitation, and submitted a report. His proposed canal connected the North Sea to the Baltic, through Lake Wenern, by a waterway 120 miles long, of which 55 miles was to be artificial. The cut was to be 42 feet wide at the bottom, 10 feet deep, with locks 120 feet by 24 feet. With the assistance of the British Government, Telford sent skilled workmen and special equipment to Sweden and, under his general guidance, the canal was brought to successful completion, remaining to this day, an economic asset of great importance to Sweden. Telford remained, for the rest of his life, on very friendly terms with the Swedes, especially the Count. His work on the Gotha Canal,

Telford's greatest canal works *were his ship canals. His work on the Gotha Canal was an early example of the invisible export of British engineering skill which developed during the 19th century.*

started as a gesture to an ally of Britain during the Napoleonic Wars, was an early example of the invisible export of British engineering skill which developed during the nineteenth century and is today of great importance to our international financial position.

The time occupied by ships on the long sea routes provides an incentive which attracts attention to possible short cuts. The long

journey around the Cape of Good Hope from London to Sydney extended to 12,600 miles and, in the days of sail, the return voyage, using the prevailing winds, was around Cape Horn, a distance of 13,350 miles. The advantages of a canal joining the Mediterranean and the Red Sea were so self-evident that it is small wonder that men studied its possibilities from ancient times. Indeed, it is possible that an irrigation ditch along the route may have been used for navigation by the ancient Egyptians. Napoleon, when he was in Egypt in 1798, ordered a survey to be made which erroneously found a difference in level of 29 feet between the two seas. In 1847, a further survey proved that the seas were, in fact, on the same level. The practicability of a canal was, in spite of this, still a subject of dispute, and Robert Stephenson, in his capacity as a Member of Parliament, advised the House that 'A canal is impossible—the thing would only be a ditch.' A French consul in Egypt, Ferdinand de Lesseps, through his friend, Said Pasha, obtained authority from the Egyptian Government to form a company for constructing a canal and ancillary works.

The Suez Canal has never at any time offered a serious technical problem to the engineer; plentiful labour was available to start the excavation by hand methods, the earth being carried from the site by basket, while the later underwater cuts were done by bucket ladder dredgers or purpose-built machines. The canal has, however, been the subject of political and financial controversy with Britain at first against the project, then, in 1875, becoming a major shareholder to join France in its control until the take-over by the Egyptians in 1956. The maintenance and enlargement of the canal has kept a fleet of dredgers continuously working, the requirements of larger ships involving the removal of nearly 600 million yards of material.

While the Suez Canal offered advantages to world shipping, the Corinth Canal, which shortened to a mere 4 miles the long trip around the Peloponnese peninsula, offered only local advantages. The ancient Greeks had a roadway with guide rails or grooves

The Suez Canal excavations *were started by hand methods, the earth being carried from the site by basket (top left). The later underwater cuts were made by bucket ladder dredgers such as that in the picture (above) of the procession at the opening in 1896.*

across which their ships could be rolled. Nero started to cut a canal, but gave it up. The modern canal was a costly undertaking as $2\frac{1}{2}$ of its 4 miles were cut through rock to a depth averaging 190 feet, the maximum being 287 feet. The work was a technical success, but at a great cost per mile and, in spite of the effort, little used by shipping.

The Kiel Canal, completed just in time for the 1914 war, was cut as a means of transferring warships quickly from the Baltic to the North Sea. It was constructed on a line partly originated by the 1784 Eider Canal, and, although the two seas are at the same level, the tidal range at the North Sea end is considerable so that sea locks are required. The canal, although inspired by war organizers, is a great commercial asset to the trade of Europe and probably carries more ships per day than any other in the world.

What the Kiel Canal did for European trade, the Manchester Ship Canal did for Lancashire. Throughout the growth of the cotton industry, the merchants of Manchester had resented the high costs of shipping their goods in and out of the port of Liverpool. Ultimately, in 1882, a prominent Manchester citizen, Daniel Adamson, called a meeting at his house to discuss the best means of connecting Manchester with the sea. A committee was formed to consider the alternative engineering possibilities, including schemes which had been proposed during the previous half-century or so. The proposal adopted was to cut a canal to Latchford with a surface level $9\frac{1}{2}$ feet above mean tide level, entered by locks from the Mersey at Eastham, to run alongside the river to Runcorn, a distance of 13 miles, then for another $7\frac{1}{2}$ miles in a straight line to Latchford, where the locks raise the level by $16\frac{1}{2}$ feet. The route then goes via Irlam where locks raise the level 16 feet, to Barton Lock with a 15-foot lift and, finally, to the Mode Wheel Lock which raises the canal by 13 feet to the level of the docks. The canal was constructed to substantial dimensions, the depth being 26 feet, with 28 feet at the lock sills; the bottom width is 170 feet for the 5 miles between Manchester and Barton, thence to Eastham 120 feet at bottom level and 172 feet at water level.

The Manchester Ship Canal was probably the most highly

Manchester is one of the great seaports *of Britain thanks to its Ship Canal. In the distance a ship is seen leaving the Runcorn Docks of the Bridgewater Canal, within sight of the transporter bridge.*

mechanized civil engineering project of the nineteenth century. The total amount of excavation was calculated to be 53½ million cubic yards, of which 12 millions were sandstone rock. Of the latter, 11½ million cubic yards were removed by excavation 'in the dry', and the remaining half million by dredging. Excavation in the dry removed 38½ million cubic yards of the softer material, while dredging cut 3 million. Quantities like these were manifestly beyond the financial resources of any company if hand labour was to be used exclusively; the canal was not being cut in a country with cheap peasant labour. Machines were therefore used on a grand scale, no less than ten dredgers and ninety-seven steam excavators being employed on the work, while, to remove the spoil to the dumping grounds, 173 locomotives and 6,300 trucks were used. The railway track laid for handling the excavated material, which included the lines laid in the canal bed, those on the embankments, the sidings, and the tracks to the spoil grounds, totalled 228 miles. This vast array of plant excavated the canal at rates varying from $\frac{3}{4}$ million to $1\frac{1}{4}$ million yards per month according to the nature of the material being handled.

The canal, unlike most of its predecessors, was not cut through primitive country, but through a highly industrialized area of the small island of Britain. The problems of its construction, although serious, were of less magnitude than those of satisfying the many interests already in being and which would be affected, or made out that they would be affected, by the construction of a canal along the proposed route. Many of these interests, where valid, could be satisfied by purely financial concessions, but many others required physical works to be undertaken for their satisfaction. Such interests included railway, road and canal undertakings, most of which called for the construction of bridges or other means of communication for the continuation of their existence; the canal is therefore crossed by a number of bridges and also, by an aqueduct to carry the Bridgewater Canal, replacing the original Barton Aqueduct of James Brindley. This is the most interesting of all the bridges crossing the ship canal and consists of a wrought-iron trough, 19 feet wide and 7 feet deep, forming the waterway. This trough is supported by a centre-pivoted swing bridge of wrought-iron open girder construction, the extreme length of the girders being 234 feet 6 inches. The aqueduct is always swung full of water as the Ship Canal Company honoured a requirement of the frontagers of the Bridgewater Canal that the water of their canal should be kept clean; it was also necessary to conserve water as summer droughts had always been a problem with Brindley's canal. The design of the end doors of the trough and of the canal abutments provided an interesting problem for the engineers.

For road crossings, swing bridges were provided at all but two of the existing roads. These were constructed to provide a head-room of 16 feet for small craft when the bridge is in the closed position. Such bridges were not unduly inconvenient to road users when the canal was built, but modern road traffic has necessitated the construction of several new high-level road bridges to carry trunk routes over the canal. With the railways, the position was different. When parliamentary powers were obtained for their construction, it was made a condition that, if the river traffic was increased by the improvement of the Mersey navigation, high-masted ships should be accommodated by the substitution of swing bridges for the fixed bridges. The canal company recognized that, with the growth of traffic during the intervening years, such a requirement was no longer tenable; they therefore provided for high-level bridges at all railway crossings of the canal. The clearance height of the bridges was decided by that of the existing high-level Runcorn Bridge, carrying the London–Liverpool line of the London and North Western Railway. In all, the canal was crossed, at the time of construction, by seven swing road bridges, a swing aqueduct, two high-level road bridges, and four deviation railway bridges.

While the Manchester Ship Canal was coming into being in Britain, men were in trouble with another ship canal on the other side of the Atlantic. The Panama project was being beaten, not by engineering difficulties, but by the mosquito. The advantages of a ship canal through the narrow isthmus joining North and South America had been realized from the time of the Spanish Con-quistadores. Many schemes had been proposed, but none had come to fruition, although the physical conditions of the country favoured such a project. In the 1870's an American scheme was being seriously considered which proposed to improve the San Juan River to Lake Nicaragua, cutting through the 28 miles of land separating the lake with the Pacific. Before any further action could materialize, a French group of financiers obtained a concession to cut a canal from Colon to Panama and floated the

A vessel travelling from the Atlantic *to the Pacific on the Panama Canal is seen entering the first of the three Gatun Locks at the northern end. The six locks on the Canal are built in pairs to allow ships to pass in both directions. The St Lawrence Seaway (right) was opened in 1959. The photograph shows one of the channels which provide a depth of 27 ft all the way from the Atlantic to Lake Superior.*

first Panama Company in 1880 with Ferdinand de Lesseps as President.

De Lesseps intended to cut a canal across the isthmus at sea level at an estimated cost of £34 million. By the end of the first five years he was overspent, his sea-level scheme was seen to be impracticable, sickness among the workmen was rampant and many of the officials were suspected of corruption. Although more money was raised in France, the company crashed into bankruptcy in 1889 with £100 million spent, only one-third of which had been actually put into the construction.

The matter caused a great scandal in France; De Lesseps was made a scapegoat and was prosecuted, receiving at eighty-eight

years of age a sentence of five years' imprisonment, which, however, he was not called on to serve. The public conscience probably realized that he did not deserve it anyway! The company
was reorganized to continue the work, but progress was slow; by
the end of the nineteenth century about 10 miles of the Atlantic
end of the canal had been dug, 4 miles of the Pacific end, substantial amounts of dredging in the entrance channels, and a small
amount of excavation at the site of the big cutting at Culebra.
In the end, following more financial and health difficulties, the
French sold their rights in 1903 to an American company
called the Isthmian Canal Commission for the absurd sum of
£8 million. The Americans tackled the engineering task with
vigour, but were soon in trouble with fever; hundreds of

labourers died and the supervisory staff would not remain on the site. The U.S. Government dismissed the Commission and appointed a new controlling body who immediately tackled the health problem with a thorough clean-up of the site, taking advantage of the new knowledge of tropical diseases which had recently become available. From then on progress was steadily made until, in April 1916, the canal was opened to admit traffic between the Atlantic and Pacific.

Its length of 40 miles saves shipping enormous distances when travelling from one coast of America to the other, a voyage from New York to San Francisco being reduced from 13,742 miles to 5,289 miles, while from Liverpool to the West Coast the voyage is shortened by 5,600 miles. Such improvements in sea transport, in addition to the safety factor involved, add greatly to the productive capacity of the countries using them, and thus, to the well-being of millions.

Modern transport calls for even more waterways. To provide for the needs of the North American continent, the St Lawrence Seaway was opened on 26 June 1959 by Queen Elizabeth II and President Eisenhower, representing the two countries, Canada and the United States, responsible for its financing, construction and operation.

In 1932 the Welland Canal was opened, joining Lake Ontario with Lake Erie by a waterway having locks 800 feet long, 80 feet wide and 30 feet over the sills. The St Lawrence Seaway is constructed to similar standards and has extended the navigable waterway for sea-going ships from Montreal to Toledo, at the western end of Lake Erie, 1,200 miles from the Atlantic. It thus links the prairies through the great lakes of Canada and the United States with the ports of the world.

*Tramways from collieries to canals — Trevithick's
steam locomotives — George Stephenson's engine
at Killingworth — the first public railway
the Stockton and Darlington Railway — the
beginning of the railway age — locomotive trials
at Rainhill — the Grand Junction Railway
Robert Stephenson's London and Birmingham Railway
the 'Parsons and Prawns' railway — Brunel and
the Great Western — George Hudson, the 'Railway
King' — atmospheric railways*

◄Kilsby Tunnel *was only one, although the greatest, of the massive
engineering problems which faced Robert Stephenson during the building
of the London–Birmingham Railway. The immensity of the operation
can be gauged from the size of this working shaft.*

Railways

THE SUCCESS OF the canal system in promoting industry led ultimately to a situation in which canals in the industrial areas of Britain were so overloaded with business that the enterprise which led to their original construction and operation gradually waned. The canal companies treated their customers with little consideration, they became selective in the goods to be taken, decided for themselves the volume of each kind of traffic and the time when it should be shipped. By the 1820's it took longer to move cotton from Liverpool to Manchester than to carry it across the Atlantic. Profiteering was rampant in the more prosperous industrial centres and charges rose to excess.

The frustration arising from these conditions, together with the ready availability of capital accruing from the prosperous industries, led manufacturers to encourage the initiation of schemes to carry goods and passengers by railway. Not only factory areas, it was believed, would gain by railways, but agriculture, especially in places not accessible to natural waterways, would gain by the quick turnover of its products. Transport was expected to be cheaper, quicker, and more reliable by railway, even before the first line was built.

By the early nineteenth century most of the conditions leading to the development of railways were capable of being met. For two centuries or more, tramways had been developed as part of the equipment of collieries and iron mines; these had been used with horse traction to connect mines over short distances with the canal system and the ports. Starting with a primitive arrangement of wooden beams and transoms, the wheels were guided by wooden battens or rails outside the wheels. At a later date, the wear was reduced by cast- or wrought-iron plates fixed to the wooden beams to form 'plate ways', from which the term

Cowpen Colliery 1618

Prior Park 1733

Whitehaven 1738

Coalbrookdale
1767

Nanpantan
Colliery 1788

Wylam
Colliery 1808

William
Jessop 1789

Lawson
Colliery 1797

'platelayer' has survived. These plates were formed with a vertical continuous flange on the outside to guide the cast-iron wheels of the wagons, this improvement having been introduced about 1767. A still later improvement consisted of turning the plate on edge, forming the cast-iron edge rail, on which ran cast-iron wheels with a flange on the inner side, the basic form of all rail wheel designs used in practice since that time.

The use of such tramways as feeder lines from colliery to canal enabled the artisan-engineers of the time to gain experience of the many developments being brought into use.

As long as the horse remained the means of power, such elementary rail systems proved reasonably satisfactory, although the increase of traffic, even by such means, called forth a steady progress in technology. The more plentiful production of wrought iron arising from the exploitation of Henry Cort's methods at Merthyr Tydfil and elsewhere led to the use of that material before the end of the eighteenth century, although without great success. Cast iron was still competitive with wrought iron when the first commercial lines were built. However, by 1820, the

A wagonway was laid in about 1730 by Ralph Allen of Prior Park to carry stone from his quarries to Bath, then being built by John Wood.

The first steam carriage was built by Richard Trevithick and tested in 1801. This original model was made by Trevithick about 1797.

problem was solved for the time with the development by John Birkinshaw, of Bedlington Ironworks, of wrought-iron rails rolled into a shape similar to the cast-iron rails. Free from the risk of breakage, the Birkinshaw rails superseded cast-iron rails on the early railways.

However good the rail tracks, horse transport alone would not have brought about the railway revolution of the nineteenth century. By 1800, however, men were thinking in terms of the new steam power, already proving its worth as a source of stationary power. In that year, Richard Trevithick started building his first full-scale road locomotive and by Christmas of the following year, he tested on a road at Camborne in Cornwall the first road carriage ever driven by steam. Public opinion was not ready for such an innovation and the Cornish genius turned his attention to rail-borne traction under the patronage of Samuel Homfray, of the Penydarren Ironworks. His successful locomotive was, however, too heavy for the rails of that works and the same fate met a second one which Trevithick built for Christopher Blackett, the owner of a colliery at Wylam, near New-

Watt's Hexagonal 1802

Surrey 1803

Le Chan 1808

Lord Carlisle 1811

John Birkinshaw 1820

castle. Trevithick's third attempt at locomotive building was made in 1808 when he laid down a small circular track near Euston Square on which he demonstrated his machine to the public at a shilling a head admission. The London public took little interest and the show soon closed and with it went Richard Trevithick's interest in steam locomotion.

Trevithick's locomotives demonstrated the practicability of steam traction; they also proved that a smooth tyre bearing on an equally smooth rail was, under sufficient weight, able to transmit the tractive effort required to draw economic loads. Trevithick also proved the advantage of higher steam pressures. He was, however, ready with his invention before a strong demand had arisen; and his volatile spirit, lacking the patience needed to pursue his invention to ultimate acceptance, took him off to an adventure in South America while other and steadier characters developed steam traction to its ultimate and complete acceptance.

Of these, John Blenkinsop in 1811 patented a locomotive with rack propulsion of which the first two, built by Matthew Murray, were put into service on the Middleton tramway in 1812, followed in 1813 by two more. These locomotives solved the immediate problem of a compromise between the weakness of the cast-iron railway, which called for a light locomotive, and the weight necessary for traction by wheel friction only. They appear to have been quite successful as the *Gentleman's Magazine* refers to them in 1825 as having given service for fourteen years.

William Hedley, of Wylam colliery, built a successful locomotive, the *Wylam Dilly* about 1815, which again proved that traction on smooth rails was only possible with stronger trackways. With Hedley the name of Timothy Hackworth is associated in the development of his machines.

George Stephenson's early days at Killingworth colliery brought him in touch with many of these early pioneers and with a born ability in things mechanical, together with a gift for observation, he was ready to take the opportunity offered to him in 1813 by Sir Charles Liddell to construct a locomotive for the Killingworth wagonway. His first locomotive, the *Blucher*, made its first trip on 25 July 1814.

Stephenson's own drawing *of the locomotive which was constructed for use on the Killingworth colliery wagonway in 1814.*

Stephenson's qualities as an engineer attracted the attention of William Losh, of the Walker Ironworks, Newcastle, and this resulted in the two men entering into a partnership to share the cost and profit of Stephenson's inventions. As a result, a great improvement was made in the design of wheels and track which led to a substantial increase in the number of Stephenson's locomotives on tramways by the 1820's.

While these Tyneside developments were going on, the Surrey Iron Railway was authorized by Parliament in 1801 as the first 'public' railway. Opened for traffic in 1805, it used horse traction and ran from Wandsworth to Croydon. The gauge was 4 feet. From 1801 to 1821, nineteen Acts passed through Parliament for the construction of railways of which fourteen were constructed.

The rich Durham coalfield suffered at that time from the lack of transport to the sea, and schemes had been considered for the construction of a canal or a tramroad. Edward Pease, a Quaker, in 1817 proposed a railway from Stockton to Darlington and through his efforts, money was subscribed for the preliminary costs of surveys and legislation. The first Bill for this railway was opposed by the influential Duke of Cleveland who alleged that the line would spoil his fox covers. Pease and his friends returned to the struggle and in April 1821 an Act was passed 'for making a Railway from the River Tees at Stockton to Wilton Park Colliery'. This Act did not envisage any form of transport other than horse-drawn wagons and its main purpose was coal traffic. In 1823 an amending Act was passed which authorized the Company to instal a stationary steam engine for rope haulage and to use locomotives or movable engines for the haulage of both passengers and goods. It was a public railway open to all who wished to put horses and wagons on it by paying the necessary tolls.

In its early discussions, the committee appointed by the shareholders of the Stockton and Darlington Railway obtained the services and advice of several engineers. They already had before them canal schemes prepared in 1768 by Robert Whitworth, the son-in-law of James Brindley, together with Brindley's own report on this proposal, another by Ralph Dodd in 1796 and a third by George Atkinson in 1800. On top of this, the committee resolved to call in the services of John Rennie who, after three years of unaccountable delay, merely repeated the scheme of Brindley and Whitworth. A financial crisis had intervened in the meantime and the proposals were delayed until 1818 when Rennie was asked to prepare further proposals in collaboration with Robert Stevenson, the Engineer of the Board of Northern Lights. Stevenson had built the Bell Rock Lighthouse for which Rennie had been the consultant and there had been some controversy over the allocation of credit for that great work. Rennie took umbrage at the suggestion that he should collaborate with Stevenson and wrote a letter to that effect to the committee which severed his connection with the scheme. In the meantime, through the influence of Thomas Meynell, the committee chairman, George Overton, an experienced tramway engineer from Wales, was engaged to make still another survey and report, and on this survey the first Bill was prepared. Overton's scheme was prepared on the basis of a horse tramroad, such roads having been

George Stephenson was convinced that steam locomotion was the transport of the future and he was engaged to prepare a survey for the Stockton and Darlington Railway.

the limit of his experience. Edward Pease, however, had by then met George Stephenson who had convinced him that steam locomotion was the transport of the future and Stephenson was engaged to prepare still another survey.

His son, Robert, had up to that time been apprenticed to Nicholas Wood, a colliery engineer friend of George, but the opportunity offered by the new survey led to Robert's release by Wood and his collaboration as assistant to his father on the Stockton and Darlington survey. Stephenson's survey and estimates were presented to the committee on 18 January 1822, and were accepted. The work was ordered to commence forthwith, George Stephenson being the engineer at a salary of £600 a year. For this sum Stephenson undertook to spend at least one week in every month on the site of the works, and also to pay his own expenses including the salaries of his assistants.

During the last stages of negotiations leading eventually to the appointment of George Stephenson as engineer for the new railway, John Birkinshaw had perfected and patented his new process for rolling double-headed rails of wrought iron in 15-foot lengths. Stephenson was enthusiastic about them and on his advice the directors adopted them for all the main running line. Their decision to retain cast-iron rails for the passing loops was evidently due entirely to the lower cost, as the wrought-iron rails of Birkinshaw's section were supplied by Longridge at £15 per ton compared with about £6 15s. od. per ton for the cast-iron rails which were supplied by the Neath Abbey Ironworks which also supplied the chairs. The cost of sea transport must have been very low to enable the South Wales industry to compete, and the ships bringing the iron must have been small, as the extensive ruins of Neath Abbey Ironworks include lengths of the old wharves, probably used for shipping these rails and chairs, and these only provide mud berths on a shallow arm of the Neath river.

The first rail of the new line was laid on 23 May 1822, and construction proceeded without delay for the next three years, and on 27 September 1825 the line was opened by the steam locomotives *Locomotion* and *Experiment* coupled in tandem to twenty-one coal wagons temporarily fitted with seats for passengers. The first journey was not without its events, but after allowing for 55 minutes' stoppages, the trip of 8½ miles was made in 65 minutes, an average of 8 miles per hour. The importance of the occasion was recognized by the many thousands of spectators and by the local worthies who dined and wined at the official banquet that night.

The Stockton and Darlington Railway marked the transition between the colliery tramroads and the railway proper; steam and horse traction were both used from the beginning, the latter predominating. From its opening, the line proved an immediate success and was extended to Wilton Park colliery, thereby doubling its length. Although coal provided most of the traffic, unexpectedly the passenger side developed steadily although in the face of keen coach competition. Ultimately the company bought out the opposition and eventually transferred all haulage to steam locomotives.

If the Stockton and Darlington is considered as representing the transitional or 'warming up' stage of railway development, the Liverpool and Manchester Railway may be considered as beginning the railway age. In 1825, a memorandum was prepared by the merchants of Liverpool who complained that

The opening of the Stockton and Darlington Railway *was
attended by large crowds who saw the steam locomotives 'Experiment'
and 'Locomotion' drawing 22 coal wagons which had been temporarily
fitted with passenger seats.*

'the merchants of this port have, for a long time past, experienced very great difficulties and obstructions in the prosecution of their business, in consequence of the high charges on the freight of goods between this town and Manchester, and of the frequent impossibility of obtaining vessels for days together'.

This state of affairs represented the inadequacy which existed in road, canal and river transport as a result of the great expansion of industry which those very transport systems had encouraged. From 1821 onwards, public opinion in the area had been veering towards the idea of a railway between Liverpool and Manchester, and in 1824 a prospectus was issued, supported by the leading merchants of the district, offering

'the establishment of a safe and cheap mode of transit for merchandise, by which the conveyance of goods between the two towns would be effected in five or six hours (instead of thirty-six hours, as by the canal) whilst the charges would be reduced by one-third'.

The carrying of passengers was envisaged in the prospectus: 'the railway holds out the fair prospect of public accommodation, the magnitude and importance of which cannot immediately be ascertained'.

This, the first railway built specifically for carrying both passengers and goods, using steam traction exclusively, naturally met with opposition from the existing interests. The canal owners saw their traffic passing to the railway and the landowners objected to the nuisance of a railway passing through or near their property. This opposition was effective at the presentation of the original Bill which came before the Commons in 1825 and was withdrawn. A modified Bill, presented in the following year, met fierce resistance, but nevertheless passed both Houses of Parliament. George Stephenson had already, in 1824, been appointed engineer for the new line, but without the assistance of his son, who had ill-advisedly accepted an invitation to lead an expedition to Colombia for the development of ports and railways to serve the mines in that country.

Without his son, Stephenson was faced with a serious task of administration in organizing the construction of a railway which included much greater problems of civil engineering than any which he had tackled in the past. In addition, the directors of the Liverpool and Manchester Railway Company had, of necessity, called in a number of engineers of professional status to assist their Bill through Parliament, and the opinions of these engineers, who included George Rennie and Josiah Jessop, often conflicted with those of Stephenson. He, in the obstinacy which was his nature, refused to accept constructive criticism which often could have been of great help, but, in other cases, he rightly rejected views which could have been detrimental to his scheme. Of the latter, Stephenson's method of crossing the bog known as Chat Moss, condemned by all the 'professional' experts, proved completely successful, although slow and costly.

On the other hand, he seemed to have lost control of the administrative side of his responsible duties and ill-advisedly undertook the main construction of the line by direct labour, acting virtually as both contractor and engineer to the project. Stephenson concentrated his first efforts entirely on the stabilization of the line over Chat Moss, which he began in June 1826, extending the activities to the rest of the line in the following January. By that time, financial difficulties led to a request for a loan of £10,000, which was granted by the Government under the powers of the Exchequer Loan Bill, for which Thomas Telford was an adviser.

This Bill was an attempt on the part of the Government to assist recovery to normal working of the nation's economy after the defeat of Napoleon. A sum of £1,750,000 was made available for subsidizing public works which, when completed, would help to accelerate the settling down process. Most of these works were to assist transport in some way or other, by constructing canals, bridges, roads and harbours.

In the latter part of 1823 another application was made to the Commissioners of the Bill for a further loan and this led to the intervention of Telford at the Commissioners' request. James Mills, an assistant to Telford, went to Liverpool in November

Timber sleepers *were used by Stephenson where the Liverpool–Manchester line was subject to settlement as at Chat Moss. Elsewhere the line was laid on stone blocks.*

1828 with Telford's instructions to inspect the progress of the works and report weekly.

Mills found Stephenson's affairs in a state of disorder and the engineer unwilling to meet him; such plans as there were he had to copy and most of the information sent to Telford was obtained by personal contact with the men at the site. To quote from one of his reports:

'There does not appear to be a single contract existing on the whole line. Stevenson [*sic*] seems to be contractor for the whole and to employ all the different people at such prices as he thinks proper to give them, the Company *finding all materials*, not only rails, and waggons but even *wheelbarrows* and planks, etc. There is some difficulty in making out the value of what is to do. . . . The first men I asked as to price said: "I have no fixed price or specified distance to take the stuff (spoil); Mr S. gives 8 pence, 10 pence or 13 pence as he thinks it deserves."

I asked him how far he was to do the cutting, he said nothing was fixed, he might go 20 yards further or half a mile. It is the same with the masonry—the Company find every material and let it from 1s. 6d. to 6s. per yard for labour.'

A succession of reports of this kind, together with the discourtesy shown by Stephenson to Mills, convinced Telford that a personal visit was necessary, so in January 1829 he went to Liverpool and, in the company of George Stephenson, made a close inspection of the progress of the works. What Telford saw there must have impressed him and, although he did not recommend the Commissioners to lend the company any more money, this was mainly due to the company's lack of policy in the matter of motive power. The choice between horses and locomotives had still to be made; stationary engines were favoured for the steeper gradients, some directors of the company even favouring this type of propulsion for the whole line. This lack of policy, together with Stephenson's unbusinesslike approach to the job, had led to delays and unnecessary expense, which Telford severely criticized in his report.

The effect of the refusal of a loan was a salutary one for the company, who put pressure on Stephenson to complete the railway in 1830, and resolved on a practical experiment to decide whether the locomotive was adequate for their needs and, if so, the best type to use. This resulted in the famous Rainhill trials, in which five locomotives, all by different makers, competed for a prize of £500 offered by the company. The trials were held at Rainhill over a course of $1\frac{1}{2}$ miles on 6 October 1829, in the presence of a vast concourse of sightseers, the judges being James Walker and R. U. Rastrick, both engineers of eminence.

Fortunately for the locomotive, Robert Stephenson had returned from Colombia in November 1827, and had from then devoted his activities to the engine building firm with which he and his father had been associated. His success was amply demonstrated at Rainhill when his engine, the famous *Rocket*, won the prize with an ample margin and, by its performance, established the steam locomotive as an adequate means of propulsion for railways.

The Rainhill success provided the impetus and encouragement needed by George Stephenson at this time and he pushed ahead on his construction works, in particular, with Chat Moss. In the face of adverse 'expert' criticism, which included an estimate by Francis Giles that the cost would be over a quarter of a million pounds, the filling was completed by the end of 1829 at a cost of £28,000. The line, including Sankey Viaduct, Edgehill Tunnel and a great cutting in rock at Olive Mount, was completed by the laying of 35 lb. per yard wrought-iron fish-bellied rails laid for the most part on stone blocks, although where the line was subject to settlement, as at Chat Moss, timber sleepers were used.

The opening of such a great work of construction was, quite rightly, made the subject of a public occasion and on 15 September 1830 eight special trains inaugurated the new line, carrying nearly 700 guests, including the Prime Minister and the Duke of Wellington, from Liverpool to Manchester. The proceedings

The famous Stephenson 'Rocket' *caused the first railway casualty when it struck Mr William Huskisson at the inauguration of the Liverpool and Manchester line in September 1830.*

were marred by an accident in which Mr William Huskisson, Member of Parliament for Liverpool, was knocked down by the *Rocket*, receiving injuries from which he died the same evening.

The new railway met with immediate success; not only was merchandise forthcoming, but passengers brought in even more revenue. At first, private coaches were allowed to travel on the line by payment of toll, but this proved unsatisfactory and the company bought them out, operating thereafter as a common carrier. Passenger traffic grew rapidly to reach half a million annually within a few years.

Proof of the practicability of steam locomotion by the Rainhill trials and its adoption by the Liverpool and Manchester Railway stimulated the promoters of new schemes all over the country. The Canterbury and Whitstable Railway, promoted in 1824 and surveyed by William James, had received parliamentary sanction. George Stephenson was invited to become the engineer, but handed over responsibility for its construction to his son Robert. This line was opened on 3 May 1830, with 4 miles of its length powered by stationary engines and rope haulage, the remaining 2 miles into Whitstable being served by one locomotive. The gradient out of Whitstable proved too much for this engine and

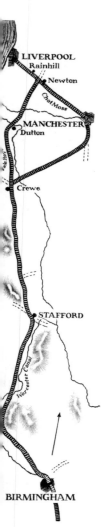

another fixed haulage plant was installed, leaving only 1 mile to the locomotive. As the line was opened for public passenger transport before the Liverpool and Manchester, it is entitled to the distinction, albeit for 2 miles only, of being the first passenger railway using steam locomotive power.

More important schemes, however, were being worked out. In 1833 Parliament approved the construction of two trunk railway lines, the Grand Junction, connecting the Liverpool and Manchester at Newton-le-Willows to Birmingham, 80 miles long, and the London and Birmingham, 112 miles long, connecting Birmingham with a London terminus at Chalk Farm, outside the suburbs. This was extended in 1835 to Euston Square.

The Grand Junction line had been planned by George Stephenson, who selected a route which would not require a gradient steeper than 1 in 330, his usual maximum. This, however, necessitated the use of tunnels, and the directors of the company decided to alter the route to one involving gradients of 1 in 260 to 1 in 180, appointing Joseph Locke, Stephenson's assistant, as their engineer. It became a characteristic of all Locke's railways that they made great use of the terrain to avoid tunnels and the more expensive works of construction. The absence on his lines of the more spectacular features of engineering had led, to some extent, to a lack of appreciation of his qualities as an engineer. The very economy and permanence of his railways entitles him to a place at least equal to any of his contemporaries.

The Grand Junction Railway, the first of many lines for which Locke was chief engineer, was also notable in being the first railway on which Thomas Brassey undertook the construction. Brassey was a land agent and surveyor from Cheshire who had worked with Telford as an assistant surveyor and who had supplied materials for the Liverpool and Manchester Railway, thus becoming acquainted with Stephenson and Locke. He tendered for part of the new line and obtained the contract for Penkridge Viaduct. His gift for the management of men and the organization of great construction works amounted to genius, and the loyalty of the navvies under his control became almost a legend. For the first time since Telford there had arisen the combination of an engineering genius with a scrupulous professional approach to his contracts and a contractor of similar technical ability in his own field on whom the engineer could rely implicitly. Brassey's railways, in due time, covered nearly 4,500 miles in many parts of the world and for many of these in Europe Joseph Locke had been responsible; the contribution of these two men to railway engineering was thus as great, if not greater, than any other such combination.

Joseph Locke obtained his early engineering training with the Stephensons, but his real genius as an engineer was inspired more by the spirit of Telford. Both men were believers in careful early surveys and planning, with specifications written for the contractor in clear, unambiguous English and they both gathered around them young engineers of good quality on whom they could rely and to whom they could delegate work of high responsibility. By such means, it was possible for a good type of contractor to tender an economic price for such works as he considered within his scope, with a reasonable chance of finishing the contract successfully at a good profit. Locke's preference for cuttings and his avoidance of tunnels, while it did not eliminate risks, reduced the hazard of the unexpected and it cannot be wondered that Brassey rose to his fame as a contractor after he hitched his wagon to Locke's ascendant star in those early railway days.

The Grand Junction Railway involved few engineering works of a nature likely to cause trouble to its chief engineer. At Penkridge, the filling of a bog was successfully achieved and the Dutton Viaduct crossing the Vale Royal near Northwich, of twenty 60-foot spans each 60 feet in height, a major engineering work, offered no difficulty. Even the most competent engineers have their weaknesses and, on this line, it was found in the failure by Locke to design a watertight aqueduct; the small cast-iron trough to carry the Bentley Canal over the railway leaked so persistently that it was completed successfully only in time for the opening in 1837.

Robert Stephenson's London and Birmingham Railway, started at the same time as the Grand Junction, was fated to run

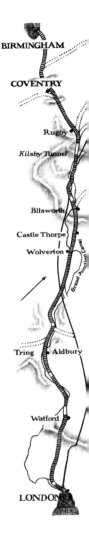

into trouble from the very beginning. Stephenson had planned with care, and had chosen first-class men for his assistant engineers; his line was divided for engineering responsibility into four divisions:

1. Camden Town to Aldbury inclusive under John Birkinshaw.
2. Tring to Castle Thorpe inclusive under John Crossley.
3. Blisworth to Kilsby Tunnel inclusive under Frank Forster, later G. H. Phipps.
4. Rugby to Birmingham inclusive under Thomas Gooch, later Frank Forster.

Each of these divisions was divided into districts with an assistant engineer in charge, who had in turn three sub-assistants to help him. The line was divided approximately into 6-mile lengths for contract purposes with special contracts for major works such as viaducts and tunnels. Many of the contractors who undertook the main contracts sublet to minor contractors who,

in many cases, profited more from the work than the often inexperienced main contractor. Such an arrangement was bound to lead to financial difficulty and failure for the men responsible for the main contracts, with consequent additional worry for Robert Stephenson.

The organization attached to such a great work, let to contractors in comparatively small pieces, was of greater magnitude than any previously required for a civil engineering undertaking. All drawings were made in triplicate, one copy each for Stephenson, his District Engineer, and the Managing Committee. For eighteen months the draughtsmen produced drawings at an average rate of one every two days each or thirty drawings per week. For this the company bought a hotel at Swiss Cottage, using the large dining-room as a main drawing office. To turn the ideas imparted by these drawings and their accompanying specifications into a railway required the efforts of a minimum of 12,000 workers, rising to 20,000 at the peak effort, on twenty-nine separate contracts let simultaneously. The administrative task was

High retaining walls *were required for the Camden Town cutting, with inverts to prevent them slipping inwards and cast-iron struts to support the tops of the walls.*

a heavy one, but through the channels organized for its performance, Stephenson managed to cope with it and even, when contractors let him down, to take over their functions by direct labour.

Although civil engineering had gained much from the experience of the great canal and road builders, a lot still remained to be learned, and the magnitude of many of the works on the new railways was bound to find gaps in the knowledge of the engineers in charge of them. Such aspects of engineering as structural theory and soil mechanics were still in the elementary stages or had not been developed at all, so it is not surprising that even the greatest of the early railway engineers made mistakes, often costly ones. Little was known of the behaviour of soil under load and particularly of the treacherous nature of London clay. High retaining walls were required for the cutting between Camden Town and Euston and these required the provision of inverts to stabilize them against slipping inwards. Primrose Hill tunnel, designed without an invert, cost more than double the estimate of £120,000 for the building of similar inverts. Embankments and cuttings also demonstrated the difficulty of supporting some materials to the slopes envisaged by the engineer.

Stephenson's greatest task on the whole line proved to be the tunnel at Kilsby. An earlier tunnel had been driven under the Kilsby ridge for the Grand Union Canal between Braunston and Crick and the survey for this canal had disclosed the presence of quicksands which had led to a realignment of the tunnel. Stephenson was aware of this risk and had, in addition, been warned by knowledgeable people of the quicksands. He ran a line of trial borings along the proposed line of the tunnel but these, unfortunately, failed to disclose a pocket of quicksand 400 yards in extent on the direct line. The contractor, James Nowell, met with difficulties through the volume of water entering the tunnel, lost his nerve, took to his bed and died of frustration. His sons failed to continue and Stephenson reorganized his forces to undertake the work directly. New shafts were sunk for pumps and the number of these was increased until ultimately thirteen pumping engines were dealing with 1,800 gallons a minute. Even with this

Kilsby Tunnel was ventilated *by means of these huge shafts sunk at regular intervals. These shafts also provided means of access for maintenance of the tunnel interior.*

plant, it was more than a year and a half before the water was beaten and the tunnel heading could continue.

Kilsby Tunnel was only one, although the greatest, of the massive engineering problems which Stephenson found it necessary to surmount during the construction of the London and Birmingham. Tunnels, viaducts, cuttings and embankments exceeding the capacity of contractors had to be taken over by the engineer for execution under his own control or that of his responsible assistants; works at Tring, Wolverton, Blisworth, Kilsby, Rugby and Coventry involving excavation and filling in unprecedented quantities cost far more in money and human life than even the most pessimistic forecast. The line ultimately cost £5½ million or £50,000 per mile whereas the estimate was £2·4 million. Joseph Locke built the Grand Junction for £18,846 per mile, and, although the terrain was much easier for Locke, it is natural to speculate whether he could have chosen a more economical alignment and organized the construction of the London and Birmingham at a much lower cost than that incurred under Stephenson.

In spite of all the difficulties, the engineers triumphed in the end, and on 24 June 1838, the first booked train ran from London to Birmingham, to be followed by regular Sunday bookings until the official opening on 17 September of the same year. The risk of slips in the embankments and cuttings led Stephenson to prohibit night trains for a period following the opening, and slips did, in fact, go on for some years. At first, trains were hauled out of Euston by a stationary engine, the gradient being too great for the locomotives available in 1838; this arrangement continued until 1844, when locomotives came to Euston and the rope haulage was discontinued.

Although the shareholders of the London and Birmingham Railway had been faced with much greater costs than were originally anticipated, they must have considered the achievement by Stephenson and his engineers well worthy of permanent record. On 3 February 1837, at their fifth half-yearly meeting, they accepted a recommendation from the directors that the London terminus should receive some architectural embellish-

Trains were at first hauled *out of Euston by a stationary engine (left). The high chimneys of the stationary engine house are seen on the right of the locomotive sheds. On the left is the eccentric for manipulating the points.*

The 'grand but simple portico' *at Euston was designed by Philip Hardwick and erected at a cost of £35,000.*

ment in the form of a 'grand but simple portico' designed by Philip Hardwick. This great monument of the Doric order, which cost the directors of the London and Birmingham Railway £35,000, has now been demolished as part of the modernization plan of British Railways.

In 1846, the London and Birmingham line was amalgamated with the Grand Junction and the Manchester and Birmingham to create the London and North Western Railway. To cope with the greatly increased traffic, the company extended the station at Euston and included in the new buildings the Great Hall. This was designed by Philip Charles Hardwick, son of the designer of the portico, whose health had broken down in 1847. The Great Hall, in an area of Euston Station planned for modernization, will eventually be sacrificed for the convenience of the large number of passengers now using Euston as an entrance to or exit from London.

While the London and Birmingham Railway was being built, other lines were being promoted and constructed in many parts of the country. Between 1825 and 1835, fifty-four Bills were approved by Parliament for the building of railways and by 1838, when the London and Birmingham was opened, railways opened for public use totalled about 500 miles. Many of these were in short lengths and unconnected to any main system and all were subject to strong opposition at all stages of their promotion. Even to survey the route of a new line required the use by the surveyors of all the guile and often all the force that they could muster.

In 1834, Bills for two great railways to serve southern England were in Parliament and the London and Southampton Act was passed in that year. The Great Western, to connect Bristol with London, had to fight its way through another session.

The London and Southampton line, popularly known at the time as the Parsons and Prawns Railway—Parsons from Winchester and Prawns from Southampton—was originally intended to include a railway and docks and the prospectus issued on 6 April 1831 proposed the raising of £1½ million of capital in shares of £25 each. Francis Giles, who had been an assistant to Telford and who was an accomplished surveyor, was appointed engineer. From the start, the railway suffered from two serious drawbacks—insufficient capital and an engineer who had not the capacity to organize the construction of a main-line railway. The financial obstacle was overcome by the intervention of W. J. Chaplin, a great man in road transport, owner of carrier carts and stage coaches. Chaplin had naturally been opposed to railways and was the strong man behind the 'Anti Rail-Road League'. A change of attitude on his part brought the London and Southampton Company the much needed capital and the benefit of an expert business man in transport affairs. Chaplin became chairman of the company in 1843 and served, with a break of one year, until 1858.

Francis Giles accepted the suggestion that he should resign and in his place Joseph Locke was appointed engineer. Locke's Grand Junction Railway was practically complete and his experience on that line enabled him to lay out a line for the London to Southampton which has given his successors little or no trouble since his time. Whereas the Stephensons, lovers of tunnelling, would probably have bored through the high chalk downs between Basingstoke and Winchester, Locke cut great slices out of the chalk and with the material raised corresponding embankments up to 90 feet high. Four small tunnels were all he required to produce a line of fair curves and reasonable gradients. It was opened on 11 May 1840, and the whole of the work was done under one contract by Thomas Brassey, who also undertook to maintain the line for ten years at an annual payment of £24,000.

Typical of the kind of opposition met by railway promoters of those days was the correspondence between Archdeacon Hoare of Winchester and Sir John Easthope, Chairman of the London and Southampton Railway. The Archdeacon regretted that the day of rest was desecrated by the running of trains on Sunday and suggested to Sir John that the divine blessing could not rest on the company in consequence. Easthope replied that 'religion would best be advanced by means of charity, and love, and kindness, which were taught, without giving unnecessary offence'. Trollope made good use of such discussions as material to enliven his 'Barchester' novels.

In 1839 the company obtained powers to build a branch to Gosport, and, in the same Act, changed its name to the London and South Western Railway. Under this title it commenced a battle for territory with the Great Western Railway Company as each extended by construction and acquisition towards the West of England.

The Great Western Railway, probably the subject of more written words than any other railway in the world, originated with a group of business men in Bristol who, on 21 January 1833, issued the following circular:

'The Gentlemen deputed by the Corporation, the Society of Merchant Venturers, the Dock Company, the Chamber of Commerce, and the Bristol and Gloucestershire Rail Road Company, to take into consideration the expediency of promoting the formation of a RAIL ROAD from BRISTOL to LONDON, request you to favour them, in writing, with such information as you may be able to afford, respecting the

expediency of the proposed Rail Road, addressed to the Chairman, in time to be laid before an adjourned meeting of the said Deputies, to be held at the COUNCIL-HOUSE on Thursday, the 31st instant, at Twelve o'Clock.

I am, etc.,

JOHN CAVE, *Chairman.*'

The Bristol and Gloucestershire Rail Road was, in fact, a horse tramroad of (non-standard) 4 foot 8 inch-gauge about 10 miles long running from the floating harbour in Bristol to collieries in the district of Coal Pit Heath. Authorized in 1828 and opened. in 1835, it was, after sundry changes by amalgamation, ultimately merged into the Midland Railway. Through this ownership, Temple Meads, the western terminus of the original Great Western, was throughout its life until nationalization a joint station.

Isambard Kingdom Brunel was invited, with three other engineers, to submit competitive proposals to the Bristol Committee; the work of surveying the line was then to be given to the entrant who estimated the lowest price. Although the temptation was great, especially to an ambitious young man not yet twenty-eight years of age, Brunel saw the impropriety of entering such a competition and informed the committee:

'You are holding out a premium to the man who will make you the most flattering promises, and it is quite obvious that he who had the least reputation at stake, or the most to gain by temporary success, and least to lose by the consequence of a disappointment, must be the winner in such a race.'

He withdrew his application if the committee enforced the requirement of the lowest estimate. By a margin of one vote his view was supported and Brunel was appointed to make the survey, the turning-point of his career. His preliminary survey followed a line through Reading, Didcot, Swindon, Chippenham, and Bath, following the Thames valley, the Wilts and Berks Canal to the Avon valley into Bristol. This scheme was adopted at a meeting of committees representing Bristol and London interests at which the name Great Western Railway was adopted

Brunel wrote on *the back of this photograph:* 'I asked Mr Lenox to stand with me, but he would not, so I alone am hung in chains.' The chains are those of the 'Great Eastern'.

and a prospectus issued. The first Parliamentary Bill was, after a long battle in committee, passed by the Commons but thrown out by the Lords. A second attempt, with a second struggle, resulted in the Great Western Railway Act receiving the Royal Assent on 31 August 1835. The Bill, by a piece of clever persuasion on the part of Brunel, was submitted without any reference to gauge, although all previous Bills had included the provision of a gauge to 4 foot 8½ inches standard, with the exception of the London and Southampton Bill in which it was omitted, apparently by accident. The broad gauge originated in a report made by Brunel to the directors of the company on 15 September 1835—only a fortnight after the Act came into being.

On this report the directors of the company resolved to adopt the 7-foot gauge for the new line. Much has been written and said since that decision on the merits and demerits of the broad gauge. Stephenson's gauge of 4 feet 8½ inches, based on the previous tramway practice, was undoubtedly on the narrow side. John Rennie advocated a gauge of 5 feet 6 inches and, if this had been adopted, there is little doubt that Brunel would never even have considered his 7-foot construction. The troubles caused in later years by break of gauge to the operating departments, by the mixed gauge to the locomotive and permanent way maintenance staffs and, finally, the vast operation of the final change of gauge must have led to many hard words, but the wide formation and generous clearances of the 7-foot gauge left the Great Western with advantages which remain to this day. The 4 feet 8½ inches gauge was standardized by Act of Parliament in 1846 for all new constructions.

George Stephenson, replying to opposition counsel while the Great Western Bill was at the committee stage in Parliament, said, 'I can imagine a better line, but I do not know of one.' This fairly assesses the magnificent route chosen for the London to Bristol line—a ruling gradient of 1 in 1,320 from Paddington to Didcot, steepening to 1 in 660 to Swindon, but with necessary gradients of 1 in 100 at Box Tunnel and Wootton Bassett.

Brunel prepared contracts which were let in September 1835 for construction at the Bristol and London ends of the line. The

The Wharncliffe Viaduct *at Hanwell carries on its south side the arms of Lord Wharncliffe, the Chairman of Committee who steered the Great Western Railway Bill through the House of Lords.*

Bristol and London committees were working independently and the character of some of the works shows some differences in emphasis on the importance of the different types of structures, those at the Bristol end having an elegance which compares with a more functional approach to design at the London end. From the London terminus, works included the Wharncliffe Viaduct at Hanwell, 300 yards long and having eight brick arches of 70 feet span 65 feet high, since widened to the same design, but still carrying on the south face the arms of Lord Wharncliffe, the Chairman of Committee who steered the Bill through the Lords. At Maidenhead, Brunel used brick again for the river crossing by a pair of brick arches, each of 128 feet span and 24 feet 6 inches

One of the most *beautiful examples of Brunel's classical architecture is the entrance to the Box Tunnel.*

rise, the longest and flattest brick arches ever built and, although their speedy collapse was forecast, carrying modern loading. Indeed, the bridge was duplicated in 1890 when the line was quadrupled. Sonning Hill was the site of a great cutting, the alternative to a tunnel which Brunel originally intended there.

So the line progressed westward, while from the Bristol end construction approached Box Hill, the site of Brunel's tunnel passing through, from east to west the great oolite, clay, the lesser oolite, blue marl and lias. After trial bores the work was begun in November 1836, and, in spite of trouble from water, was finished in the summer of 1841. The length is 3,212 yards and, surely not from coincidence, the rising sun shines through the tunnel on 9 April—Brunel's birthday. With the opening of the tunnel, the line was ready for traffic. The Bristol and Exeter Railway had by that time been constructed to Bridgwater, and to that town the first Great Western through train ran from Paddington via Bristol on 14 June 1841.

The Bristol and Exeter line was only one of many promoted during the first railway boom of 1836–7. Capital was plentiful and there was great optimism regarding the future of railways; speculators rushed to subscribe to the schemes promoted and during these two years over 1,000 miles were added to the potential of the British railway system. The new lines proposed included the London and Brighton, the Birmingham and Gloucester, the South Eastern, the Midland Counties, North Midland, York and North Midland, the Great North of England, Taff Vale, Eastern Counties, Manchester and Leeds, Glasgow and Greenock, Glasgow and Ayr. All these railways, requiring technical planning and supervision as they did, strained the resources of the engineers of the time up to and beyond the limit. The Stephensons, Locke and Brunel were in great demand as their reputations guaranteed support for any proposal with which their names were associated and it is to their credit that they withstood all attempts to engage them in the promotion of any but the soundest proposals. Brunel, incensed at attempts to entice him to associate his name with unsound schemes, even expressed the view that the term 'Consulting Engineer' meant no more than 'a man who, for a consideration, sells his name and nothing more'. He, at any rate, gave more than his name to the works with which he was associated.

'Railway King'
George Hudson

The boom of 1836–7 was followed by a period of consolidation and up to the end of 1842 a mileage of 1,857 had been completed in Britain.

The vast expansion of this period is associated in part with the career of George Hudson, who, by such corrupt practices as the payment of dividends out of capital, built up a huge railway empire through schemes of amalgamation and new promotions. Hudson's empire collapsed with the exposure of his dishonesty in 1849, but he left behind a system of unified lines, co-ordinated from the splinter systems which existed before his activities began. In fact, the railway system in Britain at the completion of this period of activity represented within a small measure that which we recognize today.

One of the consequences of this intensive period of railway construction was, through the necessity for delegation of responsibility, a great increase in the number of able young men trained and experienced in the art of railway engineering. The example of Britain in railway building was soon followed in other parts of the world, first, by the more industrial countries of Europe and by the United States who were not long in being self-sufficient in their own technology and finance of railways and, not much later, by the great overseas lands for which railways opened up such great prospects of development. In such as these, British capital and engineering found an outlet. Canada, Africa, India, Russia and South America all provided fields of investment which gave Britain the means, through the income received, to provide food and raw materials for the increasing millions of her population until, in the world wars of this century, the capital was spent to provide instead other means of survival. As an example, it is probable that most, if not all, of the railway sleepers used in Britain up to the time of the 1917 Russian revolution were paid for by interest on capital, and for goods and services supplied by Britain for the railways of Russia.

The great increase of brain-power applied to railway construction was bound to result in improvements in the technology of railways. Permanent way construction tended to become standardized, although the adventurous mind of Brunel refused

The mid-19th century *saw vast railway expansion in Britain and America. The locomotive on the left and the Niagara Suspension Bridge both belong to this period.*

to conform. His track, like all his other constructions, was intended to be more permanent and smoother running than that of his contemporaries. To this end, it was supported by longitudinal timbers resting on piles, with transoms at intervals to preserve the gauge. His rail section, unlike the double-headed section adopted by most of the engineers of his time or the flat-bottom section of Charles Vignoles, was usually a heavy bridge section screwed down to the transoms. He was, however, quite prepared to try other ideas and in South Wales he used the Barlow section, a wide based, inverted 'V' section resting directly on ballast and kept to gauge by tie-rods. His sketch-books also show beautifully detailed large scale drawings of his ideas for cast-iron chairs for use with orthodox rails on various railways in South Wales. It would appear that Brunel was quite prepared to adopt more economical methods of construction on any but his beloved West of England lines.

Brunel's most costly experiment in railway construction was the South Devon Railway. This line, from Exeter to Plymouth, included of necessity a number of heavy gradients or, alternatively, great expense in earthworks and tunnels. At the first meeting of the South Devon Railway Company, following the Act receiving the Royal Assent on 4 July 1844, a report was issued in these terms:

'Since the passing of the Act, a proposal has been received by your Directors from Messrs Samuda Brothers, the patentees of the Atmospheric Railway, to apply their system of traction to the South Devon Line. After much deliberation the Board were induced to refer the question to the Engineer, Mr Brunel, for his opinion thereon as well, in reference to the application of the principle as to the economy stated to be the consequence of its adoption. It was likewise deemed desirable that a Deputation of the Directors should visit the Atmospheric Railway now in operation from Kingstown to Dalkey, with a view to informing themselves and their colleagues of its peculiar mode of working and of the actual expenses attendant thereupon.

'From the careful consideration given to the subject by Mr Brunel, as well as from the deliberate and very decided opinion in favour of the system which he has expressed to the Board, added to the favourable report of the Deputation, and also keeping in view the fact that at many points of the line both gradients and curves will render the application of this principle particularly advantageous, *your Directors*, in the belief that it will be greatly to the interest of the Company, *have resolved that the Atmospheric System*, including the construction of an Electric Telegraph, *should be adopted upon the whole Line of the South Devon Railway.*'

The atmospheric system *at first adopted for the South Devon Railway proved to have serious disadvantages. This pumping station at Starcross, was one of a chain of such stations supplying motive power.*

The brothers Jacob and Joseph Samuda had already constructed a test track for their system and had also built the system into a small length of working line on the Dublin and Kingstown Railway. The main idea was to remove the power source from the train and to construct fixed power stations along the track at intervals. The principle had already been used extensively for drawing trains up steep inclines by rope haulage, but the Samuda patent used the pressure of the atmosphere to propel the train. A 15-inch diameter pipe laid between the rails was exhausted of air by the pumping stations and a close fitting piston in the pipe was attached to the train through a continuous slot in the top of the pipe. The slot was made airtight by a leather flap reinforced with iron which was opened automatically by the passage of the train and immediately sealed. As a means of transmitting power, the system was undoubtedly effective and the efficiency was probably higher than that of any other method then available. Unfortunately, it did not lend itself to the operation of switching the train from one line to another, but even in 1844 railway engineers looked upon traction mainly as a means of getting a train from one station to another.

127

More serious, however, were some of the other troubles to follow the opening of the line. From the first, the atmospheric pipe was liable to accumulate a large amount of water, but the first winter of use brought great difficulty with deterioration of the leather flap from various causes. Frost stiffened the leather which then cracked, the vacuum of the pipe drew out the natural oils in the leather, corrosion and chemical effects due to the interaction of the iron reinforcing plate with the leather all caused so much damage and cost so much in repair that by June 1848, Brunel reported to the directors that a complete renewal of the valve would be necessary at a cost of over £25,000. The directors then resolved to cut their losses and to run the line with locomotive power only, a loss of about £375,000. In principle, Brunel was right in his choice of separate power stations; the problem has now been solved by electrification, by far the best means of propulsion under the conditions of the South Devon Line. Brunel failed to foresee the mechanical shortcomings of the system and it is fair to wonder whether he ever asked the opinion of Daniel Gooch on the Atmospheric Line.

With the founding of the railway system of Britain on a sure basis, later developments depended on advances in technology. Of these, probably the greatest in civil engineering was through the improvements in bridge and tunnel engineering, first by the use of wrought iron by Brunel in his Chepstow and Royal Albert Bridges, and by Robert Stephenson, working with Hodgkinson and Fairbairn as a team to produce the great tubular bridges at Conway and the Menai Straits; and then by the mass production of steel due to the work of Bessemer, Siemens and Martin, the use of steel for rails and ultimately, before the end of the century, the first great bridge of steel designed by Fowler and Baker and built by Arrol across the Forth. Tunnelling techniques improved likewise, making it possible to connect South Wales to the Great Western main line by a short route through the Severn Tunnel.

IV

Early Mediterranean ports — Roman ports in Britain — London as centre of world commerce development of ports in the 18th century — fortified ports — breakwaters at Plymouth and Cherbourg The Port of London — the ports of South Wales Southampton and Portsmouth -- the Eddystone lighthouses

◄A powerful impression *of the Pool of London in the 19th century, with its ocean-going vessels and river craft, is afforded by this painting by Cole. The dome of St Paul's is seen on the skyline.*

Docks and Harbours

COMMUNICATION BY SEA dates back far beyond written history, and where ships have been used for transport, so have ports and harbours been developed for their safety and convenience. The Mediterranean, with its almost tideless coasts, lent itself naturally to the exploitation of seaborne trade, and those early traders, the Phoenicians, sailed on long journeys inside and out of their central sea in the development of their business. Some twelve centuries or more before the birth of Christ they sailed from their home port of Sidon, through the straits of Gibraltar and founded the port of Cadiz. There they built extensive warehouses to store the products of the then known world, gold, silver, copper, lead, tin and iron, obtained either by trading with native miners, or by working mines themselves. Homer, in the fifteenth book of the *Odyssey* refers to the port. As translated by E. V. Rieu in the Penguin Classics edition this reads:

> 'I come from Sidon, where they deal in bronze. I am the daughter of Arybas, and a rich man he was. But some Taphian pirates carried me off as I was coming in from the country, brought me here to this man's house and sold me. He gave a good price for me too!'

The importance of Sidon was enhanced by the proximity of the Lebanon mountains with an abundance of suitable timber for shipbuilding. The port in those days was protected by a heavy masonry mole, now destroyed by the sea, and the ancient harbour is silted up.

The port of Tyre, some 25 miles south of Sidon, included a closed harbour protected by an artificial mole. About 800 years after the port was founded, Alexander the Great, on his march to

Egypt, destroyed the city as part of his campaign to subdue Phoenicia, but he greatly improved the port and held it until the Romans took it about 300 B.C. It remained active in their hands and subsequently became a port of call for ships of Genoa, Pisa and Venice before ultimately becoming a part of the Turkish empire.

After his successful expedition to Egypt, Alexander wished to consolidate his position and, as part of this policy, founded the great port of Alexandria, linking the island of Pharos with the mainland by a great causeway. This lay within the confines of an even earlier port built by the Cretans about 1,500 years before, and which consisted of extensive artificial works of which all records and traces were lost until the remains were rediscovered in the years 1910–15. The port of Alexandria was greatly improved by the Romans, who made it a city with over half a million inhabitants, second only in importance to Rome. A feature of the port was the great lighthouse on the island of Pharos which became one of the wonders of the world.

The port of Carthage, the site of which is occupied approximately today by the city of Tunis, was the centre of a great commercial civilization which began about a century before the founding of Rome. It governed a territory extending from western Cyrenaica to the Pillars of Hercules (the straits of Gibraltar) with 300 cities, and ruled the territories of Spain, Sicily and the Mediterranean islands. The port included most elaborate works within its two harbours, including spacious quays, warehouses, arsenals and with marble porticoes at the entrance to the inner harbour. Such a great commercial asset aroused the envy of the empire builders of Rome and about 250 years B.C. there began a war for world domination which ended in 146 B.C. by the destruction of Carthage and the inclusion of its empire in that of the Romans. The ancient port is now on the one side silted up, and on the other eroded by the sea.

The city of Athens provides an early example of an out-port, that is, a port situated some distance from the city which it serves, which, in the case of Athens, was the port of Piraeus with its associated smaller ports of Munychia and Phalerum. These

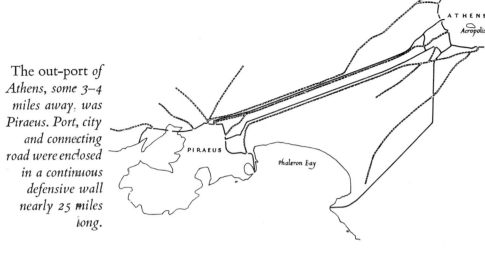

The out-port of Athens, some 3–4 miles away, was Piraeus. Port, city and connecting road were enclosed in a continuous defensive wall nearly 25 miles long.

three ports consisted of a series of natural bays and basins around the rocky isthmus of Munychia, some 3 or 4 miles from the city, the chief port, Piraeus, being on the east side and making use of a series of three natural basins. These basins were developed about 330 B.C. by the engineer Philon, who not only built warehouses and arsenals around the shores, but also sheds or roofs for 400 triremes as a protection from the weather. As the Greek ships of the time were built without decks, it is evident that a tropical storm could be a great hazard to an unattended ship in an open harbour. Not only were the port entrances protected by chains but the port, city and the connecting road were enclosed in a continuous defensive wall of nearly 25 miles length, a large part of which was of squared stones dowelled with lead. Such great works imply the use of a vast number of slaves, the Greeks themselves believing that physical work was degrading and therefore not only the labourers, but also the craftsmen of all kinds were of necessity drawn from those captured in battle or born in slavery. The port of Piraeus is today a flourishing undertaking and operates under a Port Authority which dates from 1931. Not only are the outer and inner harbours being fully developed, but the port is extending its activities to the east of the peninsula and as far as Phalerum Bay, where an even larger harbour is projected for the future.

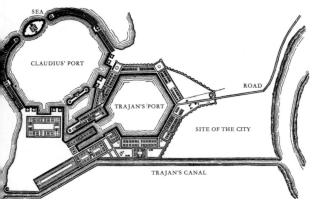

Ostia, the out-port of Rome, *was developed by Claudius and Trajan. The port of Claudius was oval in plan, consisting of two curved arms and between these was an island with a lighthouse. Trajan's port was hexagonal in shape. Roman ports were often embellished with towers and colonnades. The fresco (right) is of a harbour scene at Stabiae (now Castellammare di Stabia) at the eastern end of the Bay of Naples.*

The many ports used by the Greeks were, in the main, adaptations of nature for their purposes and generally did not involve extensive works of engineering. The Romans, however, needed safe accommodation for their ships at many centres of communication throughout their empire and, as engineers in their own right, did not hesitate to undertake civil engineering works of magnitude to provide harbours where they otherwise did not exist. Rome itself was served by Ostia, another out-port, the construction of which was begun about a century after the foundation of the city by Aneus Martius, who also built the first wooden bridge over the Tiber. The original port was successively enlarged by the emperors Augustus, Claudius and Trajan. The port of Claudius was oval in plan and almost entirely artificial, consisting of two symmetrical curved moles, each with a tower at the end, between these an oval-shaped island with matching towers at the ends and a *pharos*, or lighthouse in the centre. The whole was built of masonry and finished to a high degree of architectural decoration. The port of Trajan was hexagonal in plan and provided with extensive warehouses, shipyards and defensive works. The whole port has now silted up and is some miles from the sea.

The Romans built extensive works in masonry below water level and their success in this respect was due to their use of a cement containing volcanic ash, or pozzulana, which set rock-

hard under water. Many of the more solid under-water works must, however, have been constructed behind coffer-dams as the accuracy of the stone-laying is evidence that the work was done in the dry. There is little doubt that the Romans were capable of driving the timber piles of heavy section required for building substantial coffer-dams and it follows that their pumps were also capable of dewatering considerable volumes.

During the occupation of Britain, the Romans maintained a number of ports as a part of their communications system.

Rutupiae (Richborough), Dubris (Dover), Lemanis (Lympne), Portus Adurni (Portchester) and Clausentum (Southampton) on the south and south-east coast were on the main communication route and connected with the main road system. Other ports, such as Caerleon and Nidum (Neath) started as the means of encirclement of the resistant tribes in Wales and remained in use for the handling of the rich mineral exports made available by military success. These ports show little sign today of extensive harbour works and only the fortresses remain; the ports probably consisted of wooden quays and jetties of a more temporary nature than the prestige structures of the homeland, these latter being built largely with slave labour regardless of cost.

The end of the Roman Empire brought to an end, for many centuries, all organized trade in Europe and with it interest in the building or maintenance of docks and harbours. What ships there were used the sheltered beaches as their purpose was to raid the countries visited and a quick getaway was an essential part of the raider's tactics. Such a raid took place in A.D. 519 by Cerdic the Saxon who, according to the Anglo-Saxon Chronicle, landed at Ourd and 'on the same day fought the Welsh'. His landing-place has been the subject of much speculation, but from a tactical point of view the sheltered beach within Calshot Spit would have been ideal, with a narrow neck of land on which to defend the retreat and a gravel beach of good slope for handling the boats into sheltered water. The adjoining creek named Ower Lake, and Ower Farm nearby appear to be a corruption of Ourd.

By the twelfth and thirteenth centuries, trade in Europe was again flourishing; quays and jetties were coming into use for convenience in loading and unloading ships trading from the Mediterranean to ports in northern Europe. London, Southampton, Liverpool, Bristol, Newcastle, Shoreham and Exeter are among the ports of England which date back to those times. Not only the Mediterranean, but the Baltic ports were actively developing and these latter organized themselves with the cities of their hinterland into the powerful Hanseatic League, led by Lübeck and including some 160 towns. The league reached the peak of its influence about A.D. 1300.

The port of Southampton was typical of those enjoying special trading privileges about the twelfth to fourteenth centuries. The export of wool had reached great dimensions and, to keep the trade under control, it could not be handled except through ports chartered for the purpose and known as 'Staple' ports, this to some extent guaranteeing their prosperity. To Southampton came wool by pack-horse trains from Salisbury Plain, the Cotswolds, and elsewhere; the pack-horse men or porters used Porter's Lane for their rendezvous and for storage in the authorized wool houses there, while the horse rested in Porter's Meadow (now Queen's Park) nearby.

Pack-horse trains *were used to bring wool to the port of Southampton from Salisbury Plain and the Cotswolds.*

As these trading links between European countries developed, so the port arrangements with their structures became more elaborate and expensive. The African coast of the Mediterranean became a stronghold under the Saracens who, for their warlike purposes, constructed ports in Tunis, close to the original port of Carthage in Tangier, Gibraltar and elsewhere. Italy, as we know

137

it, was then a collection of powerful states based on successful international trade using ports such as Genoa, Pisa, Taranto, Palermo and, most important of all, Venice. Although no engineering drawings or plans exist of the port works of those days, paintings by early masters, such as Canaletto, depict harbour works which could not have developed in a decade, or even a century, and thus give some idea of the extent of early harbour engineering in the Mediterranean.

Even before the discoveries in the Western hemisphere by Columbus and his successors, the coasts of the Eastern half of the world were being linked by sea travel. Arab traders were bringing goods from India in the Middle Ages, to be followed later along the same routes by the Turks, while the Portugese navigators of Henry the Navigator worked their way down the west coast of Africa with the aid of the mariner's compass and portolan charts.

The opening up of the new world of the Western hemisphere produced a situation in which navigators no longer could creep coastwise from port to port, but instead found it necessary to provide self-sufficient ports thousands of miles from the homeland—ports in which not only goods could be stored and embarked but which could provide for the living needs of ships and even fleets in terms of food and other supplies, repairing and building facilities for ships. Sometimes, most important of all, defensive works were necessary against attack not only from local enemies, but from rival exploiters of the rich resources discovered in the new world. Such needs resulted in the founding of settlements and the development of ports not only on the coasts of the Western hemisphere, but also on the concurrently developing routes round Africa towards the East. From Lisbon, Bristol, London and Antwerp went ships to found the ports of Rio de Janeiro, Cape Town, New York, Bombay and others which offered rich rewards in goods of great demand which, in turn, were to revolutionize trade in Europe. The great and complicated pattern of commerce had, by the seventeenth century, become a matter requiring some central organization and, largely through its success in war, Britain took the reins and London became the great focal point of world commerce.

At this time, two main factors determined the commercial success of a port; one, the physical shape of the harbour and its suitability for shipping and the other, the capacity of its immediate hinterland for the production or absorption of goods carried by sea. It is mainly by changes in these factors that ports have risen or fallen in importance. The suitability of a port for ships may vary through natural causes and with the development of the ships themselves. Steam propulsion, for instance, has made it possible to develop ports quite unsuitable for sail. The coming of railways opened up vast areas of hinterland and made it economically possible to provide artificial works, to dredge and to develop ports of far greater complication than would have been previously within the capacity of a smaller area of distribution.

The suitability for shipping of a harbour may remain permanent, although its importance as an economic asset may change; on the other hand, many ports have so changed in character through siltation, littoral drift, or other causes, that their maintenance may be a costly burden on an economy, and this may lead to their abandonment. In some cases, the strategic position of a . harbour may justify great expense by a nation, as in the cases of Plymouth and Cherbourg, both of which were of such importance in Napoleonic times that great breakwaters were built for their improvement as sheltering places for large fleets. Into a similar category come the harbours which are dependent on large artificial works such as moles or arms; these may be established at great expense where no natural features exist, but where other circumstances pertain, such as the proximity of trade routes, an industrial hinterland or, in the past, the need for a refuge harbour for storm-bound ships. All such needs have been met by great works of civil engineering and while ships are used for commerce, so will such works be undertaken.

Dock systems usually represent an adaptation of local conditions to suit developing trade. As soon as ships became too large for beaching, their cargoes were either transhipped into smaller craft to be brought ashore, or the shore line was artificially extended by some kind of pier or quay to deep water. Tidal ranges, if too great, led to the increasing use of impounded or wet docks with lock systems and the demand for berthing space for more ships led to the design of dock systems of sometimes complicated shape to provide long runs of dock or wharf wall for the purpose. From simple piled timber or timber and stone piers, engineers developed docks by the diversion of rivers, by excavation, reclamation or the tipping of heavy embankments.

Marshy areas have, in many instances, been exploited for the establishment of docks, as, for instance, the low-lying lands enclosed by the sinuous Thames as it passes towards the sea after flowing under London Bridge.

The entire docks system of Southampton represents progressive winning from the mudlands on the foreshore of the peninsula between the rivers Test and Itchen, and it is safe to say that every square foot of Southampton Docks has been reclaimed in this way. The extensive Itchen wharves, independently owned and built, are also almost entirely reclaimed from the river. Bristol, however, owes its present existence as a port to the by-passing of the rivers Avon and Frome by a 'new cut'. Many small fishing harbours are protected by massive works of masonry built in days when the labour for their construction was plentiful and cheap, and when fishing was a lucrative trade (including the smuggling). Such ports are nowadays frequently in serious financial difficulty through the changed pattern of the economy rendering repairs to their massive walls prohibitively expensive.

The Renaissance and its almost inevitable consequence, the Industrial Revolution, brought about a great development of

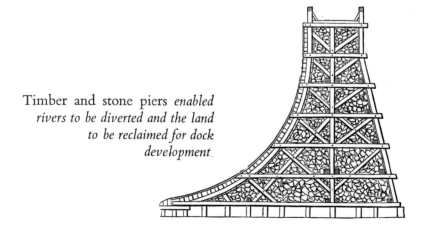

Timber and stone piers *enabled rivers to be diverted and the land to be reclaimed for dock development.*

The famous diarist Samuel Pepys *records on 15 January 1661 '. . . we took barge and went to Blackwall and viewed the dock, and the new West Dock, which is newly made there'. Blackwall was the earliest wet dock in Britain.*

ports, and this became extremely rapid in the eighteenth century. In London, Chatham and Portsmouth, the competence of Samuel Pepys as Secretary of the Navy had already established works for the Navy which were to have a profound influence on history in the next two centuries. At Blackwall there had been built the earliest wet dock in Britain, a small fitting-out basin of $1\frac{1}{2}$ acres with lock gates and on 15 January 1661, Pepys records in his diary:

> 'So after a cupp of burnt wine at the taverne there (*Deptford*) we took barge and went to Blackwall, and viewed the dock, and the new West Dock, which is newly made there, and a brave new merchantman which is to be launched shortly, and they say to be called the Royal Oake.'

At Rotherhithe, in 1703, the Howland Great Wet Dock, 12 acres in extent, was built by private capital. The lock entrance was 150 feet long, 44 feet wide, and 17 feet deep at high water, ordinary spring tides; thus being large enough for any merchant ship of the time. The dock walls were of timber construction and lasted in that form until the dock was bought by a local ship-builder called Wells, who rebuilt it under the name of Greenland Dock for the whaling trade; it changed hands again in 1806 and as Commercial Dock was used for the timber trade. In 1808 it was reconstructed to be 1,000 feet long, 450 feet wide and 18½ feet deep below H.W.O.S.T. The walls were built of timber and lime-concrete while the lock was rebuilt to 209 feet long, 42 feet wide and 18½ feet deep, the section being elliptical. It now forms part of the Surrey Commercial Docks system.

The great interest in civil engineering shown by the French from the seventeenth century onwards had its effect on their ports. At Le Havre, extensive fortifications were built by François I from 1509 onwards, as a defence against Henry VIII. The harbour of Le Havre at that time sheltered a fleet of 135 ships intended to oppose the English attacks on Boulogne. Work was done in the port by Vauban in the seventeenth century and later, following the Revolution, de Cessart was employed to rebuild the locks and effect other improvements. The same engineer also rebuilt the quay at Rouen, which had become too narrow for the greatly increased trade. The new quay wall was 120 feet in advance of the original wall and was supported on a timber platform carried on timber piles, the whole being stabilized by a tipped bank which on the land side was brought up to quay level to reclaim the area required.

In 1715, the port of Dunkirk was improved by the great French engineer Belidor, who built locks to provide wet dock facilities. This port has remained important to the present day; serving, as it does, the highly industrialized Pas de Calais area of Northern France. It is, however, of local importance compared with Le Havre, which, from its position on the mouth of the Seine, serves not only Paris but also the whole of France with many commodities.

The port of Liverpool, granted a charter by King John in 1207 when it possessed only a creek or 'pool' off the Mersey, began a new development with the opening in 1720 of its first dock, 650 feet by 250 feet, to be followed in 1753 by the Salthouse dock, 640 feet long and 306 feet wide, and St George's Dock, 750 feet by 300 feet, in 1762. These docks, with their successors, form a riverside system which expanded greatly in the first half of the nineteenth century and which with the growth of shipping has

Plans for the fortification of Dunkirk and its port were drawn up by Vauban about 1680.

required constant improvement of its approaches from the sea. In spite of this, by costly maintenance, the position of the port of Liverpool in relation to the industrial areas of the north Midlands and Lancashire enabled it to become and to be today one of the world's greatest ports.

The need for safe harbours for the fleets of Britain and France, then in opposition, led to the almost simultaneous construction of breakwaters at Plymouth and Cherbourg. The harbour of Plymouth, with its inner harbours of the Cattewater and the Hamoaze, possessed natural depths of water of 4 to 20 fathoms at low water of spring tides. Situated as it is, opposite the French ports of Cherbourg and Le Havre, it was, during the French wars, of great strategic importance. Unfortunately, the Sound, offering about 4,000 acres of anchorage, lay open to the south, and this had on many occasions led to large ships dragging their moorings and being driven ashore. The provision of a shield against these conditions had long been under consideration by the Admiralty, but the pressure of war conditions led their Lordships to invite John Rennie to make recommendations and this materialized in a report presented by him on 22 April 1806. In his report Rennie recommended the raising of a breakwater of rubble blocks tipped at random and weighing from 2 to 10 tons each; these were to form an embankment 5,100 feet long, with the central 3,000 feet straight across the entrance to the Sound, and 1,050 feet at each end turning 20 degrees inwards towards the land. He also recommended a double-armed pier, 2,400 feet long, at the south-east entrance, the whole scheme being estimated to cost £1,102,440. The scheme met with much opposition and for five years all the arguments for and against were discussed; and at last, the Admiralty decided to adopt Rennie's proposal for the breakwater, but not the south-east pier. An Order in Council dated 22 June 1806 authorized the work which was started on a basis of contracts for labour only, the Government providing all materials and equipment. Limestone was quarried at Oreston from 25 acres of land bought from the Duke of Bedford.

Work started on 12 August 1811, Joseph Whidley being the resident engineer. By March 1813, the first stones appeared at low

Rennie's breakwater at Plymouth, *completed in 1848, consisted of limestone rubble tipped at random and faced with a layer of fitted masonry blocks.*

water; by July the exposed portion extended for more than 2,000 feet and was already providing appreciable shelter. In 1815 the Admiralty, on Rennie's advice, resolved to raise the top level 20 feet above low water, spring tides, instead of the 10 feet originally intended and by 1816 the blocks were being deposited at a rate of 1,030 tons per day. The whole operation by this time had become highly organized, using plant and methods which were very advanced for the time. January 1817 brought a series of gales which moved many of the stones already deposited, some of which, weighing 2 to 5 tons, were thrown from the seaward face to the harbour side of the breakwater. The seaward slope was flattened from 3 to 1 to an average slope of 5½ to 1, but in spite of this the breakwater was proved to be effective as no ship under its protection was lost; two which had anchored outside its protection were lost by driving ashore.

John Rennie died on 4 October 1821, and the work continued under his sons John and George, who completed it in 1848. By that time, 3,670,444 tons of rubble had been cut from the quarry at Oreston, shipped to the breakwater site and tipped at a cost varying from 2s. 9d. to 1s. per ton. On this random mound, a substantial face of fitted masonry blocks had been laid, totalling 22,149 cubic yards and the whole cost of the work was about £1½ million, or less than 50 per cent more than that estimated by Rennie over forty years earlier. Truly a magnificent achievement, it stands today as firmly as it did over a century ago.

Rennie's breakwater, while still under construction, was admired by Napoleon when, in 1815, he entered Plymouth Sound a prisoner on the *Bellerophon*. The admiration came from a knowledgeable critic, as, on the opposite side of the English Channel, Napoleon had seen the development at Cherbourg of a great breakwater designed by the engineer Louis Alexander de Cessart.

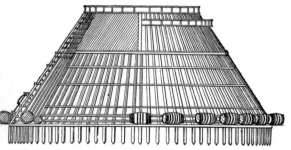

The 18th-century breakwater at Cherbourg and the 20th-century Mulberry harbour (right) employed the same principle of construction—floating prefabricated units into position and sinking them on to the sea bed. The Cherbourg cones (left) were floated by casks attached by rope.

Commenced in 1783, this work was of more elaborate design, although not as extensive as Rennie's later work, its construction being of necessity following the destruction of the port of Dunkirk. Unlike the Plymouth breakwater, which was only for defence against the sea, that of Cherbourg was intended to provide an addition for defence against an enemy and its building was therefore a matter of urgency. De Cessart chose a method of construction which was the same in principle as that adopted for the Mulberry harbour built on the nearby coast over a century and a half later, i.e. of prefabricating units elsewhere, floating them into position, and sinking them on to the sea bed. The original proposal had been to construct ninety cones, each 150 feet diameter at the bottom, 60 feet diameter at the truncated top, and 70 feet high. These were to be laid close alongside each other across the open entrance of the harbour, but evidently the demands on material and manpower could not be spared by the French shipbuilding industry and the scheme as executed included only eighteen cones; although twenty-one were built, three were broken up and sold during the Revolution.

The conical towers were built by shipwrights of squared oak and beech timbers, 13 inches square at the bottom reducing to 8 inches square at the top, with horizontal walings at intervals and two platforms at the top for use in the sinking operations. They were ballasted with enough stone to sink them, and floated into position with huge casks attached by rope ties to the base of the cone. After being towed into position, the towers were sunk by

cutting the casks free; this was done with a guillotine type of knife, weighted with lead. The eighteen cones were placed at intervals varying from 159 feet to 1,845 feet over a distance of 12,470 feet and leaving a wide entrance to the harbour at both ends; they were then filled with stones and the intervals between them made up with tipped stone. By 1795, a total of 100 million cubic feet of stone had been tipped. The work cost about a million pounds and it is evident that, even at the prices of the day, a great deal of conscripted labour must have been used. The cones were soon broken up by the forces of nature, none lasting more than four years. No doubt the marine borers, the gribble and teredo had a lot to do with this. The stones, however, fell to their natural slope and provided a great measure of shelter for the port.

The Port of London's dock system is associated with many great names in civil engineering, especially during the expansion years of the nineteenth century. William Jessop was the engineer responsible for the construction of the West India Docks, for a company of West India merchants, between 1799 and 1802 on the Isle of Dogs, the architect for the warehouses being Mr Gwilt. As the name indicates, they were built for the West Indies trade and consisted of two docks, the Import, 2,600 feet long and 500 feet wide, and the Export, which was the same length, but 100 feet narrower, the depth of both being 24 feet at L.W.O.S.T. The docks communicated with the river through a basin at each end, common to both. At the eastern end, the Blackwall Basin, oval-shaped and $6\frac{3}{4}$ acres in extent, was used for shipping, while at the western end a smaller basin of $1\frac{1}{4}$ acres, known as the Limehouse Basin, was used by lighters. Locks separated the basins from the docks and the river. The walls were founded on the gravel and were curved in section, with counterforts at the back; such curved dock walls were common at this time and appear to have been based on the assumption that the curved shape of a clay bank was the most natural shape to adopt. Due to the sound workmanship and the good foundations, the walls are still as good today as when they were built. This dock system was extended in the 1860's by the construction of the South Dock by John Hawkshaw, with Vernon-Harcourt as resident engineer.

Another company built the London Docks progressively at Wapping and Shadwell on ground previously occupied by slum property. The engineer was John Rennie and Alexander was the architect. Between 1802 and 1805 they built the Western Dock, 1,260 feet by 690 feet, and its 3-acre basin, both 20 feet below Trinity High Water, the latter being an arbitrary datum, made necessary because of the variation of high water level at different parts of the river. The walls of Western Dock are of brick similar to those of West India Docks and were built on beech and elm planking laid on the gravel, with timber sheet piling supporting the toes. The lock was constructed behind a coffer-dam of timber piles which were driven with the aid of an 8-horsepower steam winch, the first recorded use of steam for such a purpose. Steam was also used for pumping in the form of an engine of 25 horsepower by Boulton and Watt. In 1815, the Hermitage Basin and the Tobacco Dock, of similar construction, were added and in 1832 the 'old' Shadwell Basin and locks were constructed by H. R. Palmer, the founder of the Institution of Civil Engineers. Between 1854 and 1858, James Meadows Rendel built the 'new' Shadwell Basin on clay as the depth required (28 feet below T.H.W.) was below the gravel stratum.

The burst of dock building enterprise in London continued with the building, by the East India Dock Company at Blackwall, of a considerable system of extensions to the Brunswick Dock. This latter, 8 acres in extent, was built in 1789 by Mr Perry, a shipbuilder, for the same company and was used for the repair and fitting out of the company's ships. Between 1803 and 1806, the company built a dock 1,410 feet by 560 feet which was called the Import Dock, connecting it with the Brunswick Dock which they re-named Export Dock. The wall design of the new dock was similar to that of the West India Docks and the locks were adequate to take the company's largest vessels of 1,500 tons burthen. Unfortunately, the dock walls had of necessity to be founded on clay, below the ballast stratum, and in 1879 the south wall of the Import Dock slipped forward and subsided. It was later repaired and stabilized with piling and mass concrete work in front. In 1943, the same dock was dried out for the construction of 'Phoenix' units for the Mulberry harbours and similar slips occurred to the west and part of the south wall.

Thomas Telford, who had already been engaged in harbour works at Aberdeen, was consulted by still another company who proposed to build docks on a site immediately east of the Tower of London belonging to the Hospital of St Katherine, a religious order founded in 1148. This site was occupied by the Church of St Katherine, built in 1443, and by a great number of slum dwellings, so the scheme aroused great opposition. This new company, of some of the principal merchants in the city, arose from objections to the excessive charges by the existing dock companies and the hope of savings accruing from the ownership of their own docks. This, together with its proximity to the city, led them to choose a site which set a serious problem to the engineer. To quote Telford himself:

'The whole extent is no more than twenty-seven acres of a very irregular figure, so that, when the space necessary for warehouses and entrances was subtracted, ten acres only remained for the actual docks, which therefore required an unusual arrangement so as to provide wharfage and quays for the various branches of trade expected to frequent them. It being obvious that the accommodation required could not be obtained by the simple forms of squares and parallelograms, I was, from necessity, led to adapt the shape of the docks to that of the ground; and this was so managed, after attentive considerations, as to become really advantageous, as affording an increased extent of wharfage, and two docks instead of one, by which distribution the trade was likely to be better arranged; with a further advantage, that in case it should at any time be found necessary to empty one dock, the water may be retained at full height in the other.'

To maintain the water level of the dock, steam pumps driven by two engines of 80 horsepower each were provided; these were built, to use Telford's words, by 'my friend, Mr James Watt, and his able and ingenious assistant, Mr Murdoch.'

The Act of Parliament for St Katherine's Dock was passed in

The clearing of the site *of St Katherine's Dock to the east of the Tower of London involved the demolition of a church and a large number of slum dwellings.*

June 1825, the western dock, basin, and entrance lock with their associated warehouses were completed and in use on 25 October 1828, and, exactly one year later, the eastern dock was opened to traffic. Telford's resident engineer on this work was Thomas Rhodes, who had been his master carpenter for many years and to whom must be given much of the credit for the excellence of St Katherine's Dock. Telford criticized the haste with which the work was pursued when, in spite of the successful conclusion of the work, he wrote:

Huge crowds lined the quays *at the official opening in October 1828 of the western of the two St Katherine's Docks in London. The 'Elizabeth'*

and 'Mary' were dressed overall in honour of the occasion. The second
dock was opened to traffic exactly one year later.

'I must be allowed to protest against such haste, pregnant, as it was, and ever will be, with risks, which, in more instances than one, severely taxed all my experience and skill, involving dangerously the reputation of the directors and their engineer.'

The new docks were a commercial success, although there was competition by the other dock companies and, in 1830, a total of 893 ships used the dock with goods totalling 141,751 tons, bringing the directors a profit of £50,351 on their capital of £600,000.

The Surrey Commercial Docks originated in 1801 as a proposal to build a canal with docks from the Thames near the present Surrey Entrance lock, towards Deptford, thence to Camberwell, back to the river at Vauxhall, ultimately to Croydon and thence to Portsmouth. The canal to Camberwell and a branch to Peckham were constructed, but the dock interests predominated and led to the addition of a number of timber ponds, together with the Stave and Russia Docks. These docks were amalgamated in 1867 with the system based on the development of the Howland Great Wet Dock into a company called the Surrey Commercial Docks Company and were further extended in 1876 by the construction of the Canada Dock. Further reconstruction in the 1890s greatly improved the system.

The steady growth of trade in the Port of London led to the formation of still more dock companies, with greater docks extending further down the river, enabling larger ships to use the port. Victoria Docks was constructed in 1854 by G. P. Bidder, who used cast-iron piles of T section with cast-iron plates between them as part of his lock wall construction. He also fell into the error, common to most engineers of the time, of removing the gravel stratum at the dock bottom and replacing it with puddled clay on the assumption that the dock would lose water through the pervious ballast. Unfortunately, the puddled floor was lifted by water pressure, allowing the dock walls to subside and slide forward. Repairs involved the deeper driving of piles, the rebuilding of the walls, and the replacing of the puddle clay with lime concrete. This dock was the first to use hydraulic machinery for

gate, sluice, bridge, and crane operation. In 1864, a company was
formed which led to the construction of Millwall Docks, for
which John Fowler and William Wilson were the engineers.
This dock incorporated the first dry dock in one of London's
impounded docks.

The success of the Royal Victoria Dock led, in 1864, to the
amalgamation of the London, St Katherine and Victoria Dock
Companies and to the construction by the new company between
1874 and 1880 of the Royal Albert Dock, for which the engineer
was Alexander Meadows Rendel. This dock was, when it was
built, the longest in the world, being 6,600 feet long and 490
feet wide, with a depth of 27 feet below T.H.W. It connected
at one end by an 80-foot-wide passage with the Royal Victoria
Dock, through a pair of lock gates to a tidal basin and thence to a
lock entrance 550 feet by 80 feet and 30 feet deep. Portland cement
concrete was used for the dock walls, the first use of this material

Brunswick Docks on the Thames at Blackwall, intended for the
East India Company, were built to the design and at the expense of
John Perry.

for such purposes in the port. The rapid increase in the size of ships about this time was demonstrated by the necessity for a second lock, built in 1887 alongside the first and to a similar design but 36 feet deep, which was adequate to take the largest ships using the port until 1903. The Royal group of docks was further extended in 1912 by the King George V Dock, 4,500 feet long, 630 feet wide, tapering to 430 feet and 38 feet deep.

The expansion of facilities achieved by the London and St Katherine Docks Company was not equalled by the rival East and West India Docks Company. The latter had well positioned docks, extensive warehouses and good berthing accommodation, but suffered from defects of railway access and inadequate entrance locks; in consequence the trade in these docks declined. The company, in an endeavour to recapture the business, embarked on the construction of a new dock system at Tilbury which was opened in 1886. The estimated cost was £1,100,000, but, owing mainly to engineering difficulties in the excavation of the soft material, the ultimate cost was £2,800,000 and the result disappointing. The design of the docks offered difficulties in handling the ships for which adequate depths had been provided and the tidal basin proved to be a trap for mud which has been a constant expense for dredging. Not only that, but the shipping still preferred the docks nearer the heart of London. Tilbury Docks proved to be, to some extent, a white elephant, until the construction of a new upper entrance in 1928, which, with extensions to the dock carried out soon after the Authority was constituted, made Tilbury capable of handling any ship normally using the Port of London.

The Port of London today is administered under a public authority, created by Act of Parliament in 1909, which vests in it the whole management of the port including the river below Teddington.

The vast population of London assures its port of steady and, indeed, increasing business, and any port serving a large population by general trade is, because of the need for regular supplies of food and raw materials, cushioned against changes in the country's economic pattern. Liverpool and Manchester continue as major

Barges manœuvre into position *for the sea-going vessels to unload their cargoes in this 19th-century painting of the Port of London.*

ports in spite of vicissitudes in the cotton industry and Southampton maintains its position as an out-port of London. Ports dependent on specialized trade are not so fortunate as, if the local industry declines, the trade of the port declines with it. This has happened in recent years to the South Wales group of ports, which depend mainly on the coal trade which has been steadily declining since the First World War.

The earliest ports in South Wales to develop were those associated with the metallurgical industries which have been an incentive to trade since Phoenician times. The monastic houses of Neath and Margam developed the exploitation of non-ferrous metals in the Swansea Bay area and their activities were continued by private enterprise, so that by the end of the eighteenth century, the copper works of the Swansea district were supplying a large proportion of all the copper produced in Great Britain. By that time, the demand for coal was rapidly increasing and with the opening up of canals in the mining valleys came a need for rapid port expansion. In 1791, the Swansea Harbour Trust Act was passed, enabling that body to execute works in the port, but these were confined to the improvement of river wharves and the trustees did not construct their first dock until sixty years later, when the North Dock, a length of the river Tawe, was impounded and by-passed by a 'New Cut'. The South Dock, on the foreshore west of the river, was opened in 1859 and the Prince of Wales Dock on the east foreshore in 1881; both of these docks were of cut and fill construction, the excavated material making up the level of the surrounding land for the building of sheds and railways. The King's and Queen's Docks were, in 1909, added to the seaward side of the Prince of Wales Dock by excavation behind a large sea embankment. As with all the Bristol Channel ports, the great range of tide necessitates, at Swansea, the use of impounded docks. These were, for many years, fed with fresh water from local streams, but the catchment area being small and summer drought causing difficulties, the dock level is now ensured by pumping.

The nearby port of Port Talbot, thus named from 1837 onwards, began as landing places in the estuary of the river Afon; these were developed at first by the monastic house of Margam Abbey and, from the founding of that establishment in 1147 to the Dissolution in 1537, the mineral trade was fostered by the Abbey. From that time, the copper industry developed by private enterprise and by the early nineteenth century the 'Company of Copper Miners of England' had become firmly established in the Afon valley, particularly at Cwmafon. The importance of this trade led to the formation of a company which, in 1833, obtained an Act for the diversion of the river. In 1836 this was done, the forces of nature being used for the purpose by damming up the original channel, cutting a new leader channel during the dry season, and waiting for the winter floods to scour the new cut. The trade of the port increased by the export of coal, which, from the 1850s, the South Wales Railway and the South Wales Mineral Railway brought to the port, followed in 1885 by a still

A port since earliest times, *the importance of Cardiff increased with the development of canals and railways.*

greater quantity brought in by the Rhondda and Swansea Bay Railway. This increased trade led to the construction of a new dock in 1897, by which time the trade of the port was mainly in coal. In 1905, a new factor entered the economy of the port with the opening of the Port Talbot steel works, and from that time the bias of trade has changed from coal to steel with the steady rise of the steel industry. The gradual change from copper to coal and then from coal to steel has thus substantially maintained the prosperity of the port throughout its existence.

Cardiff has been a port since the earliest times, but its rise to importance came with the development, first of the Glamorgan Canal which linked it with the iron and coal industries of the Merthyr Tydfil district, and later with the development of railways in the mining valleys. The second Marquis of Bute was responsible for the development, in 1835, of the West Bute Dock and this was followed in 1855 by the East Bute Dock. Further extensions were the Roath Basin in 1875, Roath Dock in 1885 and the Queen Alexandra Dock in 1907, the last new dock to be built at Cardiff. The nearby dock and harbour of Penarth were built to serve the vast exports of coal brought down by the Taff Vale Railway and in 1899, the coal owners, who had formed the Barry Docks and Railway Company, opened docks at Barry to relieve the congestion in the port of Cardiff. All these ports grew up on the coal trade, and little provision was made for handling other traffic; with the falling off of coal exports after 1914, they suffered a great loss of business. In Cardiff, only four coaling appliances remained in use in 1961 out of a total of thirty once operating. Penarth is closed to traffic, while Barry is searching for business and has recently opened up small imports of bananas. Even in shipping, habit is an important factor; a great group of ports has now the task of creating a new outlook in the minds of shippers in the Midlands and in their own valleys a light industry is growing up which still uses ports less conveniently placed for the export of its products.

The port of Newport, most easterly of the South Wales group, on the estuary of the river Usk was first organized commercially in 1836 by the creation of the Newport Harbour Commissioners, who regulated traffic at the river wharves. Dock construction began in 1842 and continued until 1914 when the present lock entrance was opened. Newport, unlike the Cardiff group of docks, has always maintained accommodation for general cargo in addition to its considerable coal traffic. This has enabled the port to continue as an economic business and its future appears promising with the great expansion near by of the steel industry.

A minor port in South Wales is of interest as being one of the few dock constructions undertaken by Isambard Kingdom Brunel; it is Briton Ferry, associated with the South Wales Railway designed by Brunel, but completed after his death by Robert Pearson Brereton, his assistant, and opened in 1861. Although Brunel's docks here, at Millbay Dock, Plymouth, and at Brent-

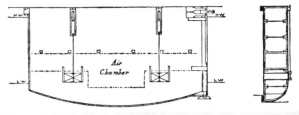

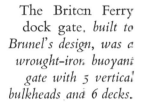

The Briton Ferry
dock gate, *built to
Brunel's design, was a
wrought-iron buoyant
gate with 5 vertical
bulkheads and 6 decks.*

ford are not works of great magnitude, the wrought-iron gates
of his design are of interest and led to improvements in the
design of semi-buoyant dock gates. That at Briton Ferry is out of
use, the dock now being a tidal basin and little used. The author
had the privilege of working on the gate when it was last repaired
in the year 1929.

If London is a port serving that great city, the South Wales
group providing the means of trade for specialized industries,
Southampton is mainly dependent on the movement of people.
Its natural advantages lend themselves to the movement of great
ships. It is a drowned river estuary sheltered by the Isle of Wight,
with a tidal régime which might almost have been made to fit the
business of the port. A moderate range of tide, 13 feet at springs,
8 feet at neaps, is accompanied by a double high water phenomenon with two hours between peaks, thus giving three hours
of high water virtually free of tidal currents. The flood tide is
also blessed with a slack period of about an hour at half tide
known as the 'young flood' which again greatly assists the move-

ment of ships. This tidal characteristic is probably the real source of the Canute legend which, in the course of centuries and the writings of ecclesiastics unversed in the sea, became distorted into the well-known story of the flattering courtiers and the relentless sea. It is more probable that Canute, king of a sea-going people, being told of the favourable but unusual tides at this place, sat on the shore to observe for himself the curious phenomenon. Long before Canute, Bronze Age traders came to the port and their products are still being found on its shores; after them the Romans occupied a fortified site, Clausentum on the River Itchen, where lead from the Mendips was shipped to the home country and from which port they could transfer goods to Venta Belgarum (Winchester) by the navigable Itchen. The Saxons occupied a site on low marshlands further down the river but the Normans eventually began the development of the peninsula between the rivers Test and Itchen which has continued, even if spasmodically at first, until today.

The trade of the port with the Mediterranean, begun already before the Normans came, grew following their occupation through the natural links with the homeland ports across the Channel and with it the town grew into a compact layout surrounded by defensive walls. Here, under their protection, lay the galleys of Genoa, Florence and Venice, to unload their cargoes of fine fabrics, dried fruits, dyes, the wines and spices of the sunlit countries. In return they took the wool of the downlands of England. These goods were landed and embarked, at first over hards or 'hithes', then across wooden jetties of piled construction and, later, at one of the three 'Keys'; the West Quay, the Watergate Quay and the Castle Quay, the last being for royal business only. These quays were of solid filling behind timber piled and sheeted or masonry walls, and represented the beginning of an encroachment on the mudlands which has continued until the present day.

The sixteenth century brought legislation prohibiting the export of raw wool for the encouragement of English wool cloth manufacture; it also brought the suppression of the monastic institutions including those of Netley, Romsey, Winchester and Beaulieu, all of which had contributed to the trade of the port, which from that time declined. By the end of the eighteenth century, Southampton's main fame was as a minor spa, but there had developed in the town a hard core of business men who were no longer satisfied to accept things as they were, and by their exertions, against the usual opposition, an Act was passed in 1803 vesting the management of the Harbour in a body of Commissioners. Although the Act by its resounding title appeared to give the Commissioners ample powers, its financial and other limitations prevented the intended great works being undertaken. The Commissioners did, however, engage John Rennie to survey the port and make proposals for its improvement. This report was referred to by Samuel Smiles in his life of Rennie as being of great importance, but until recently, when a copy was discovered and acquired by Southampton Corporation, its real significance had not been realized. Rennie's recommendations, although not adhered to in detail, were, in fact, a wonderful forecast of the future development of Southampton up to the end of the nineteenth century. The Royal Pier was built in 1833 to provide access at all tides for the steam packets operating from the port to the Isle of Wight and the near Continental ports. It was also envisaged that it would serve the P & O steamers and those of the Royal Mail Company when they came to the port, but fortunately by that time more substantial facilities were available for them. The first structure of the Royal Pier, designed by John Doswell, the engineer to the Harbour Commissioners, was of Memel timber, and as in those days no cheap method of protection had been devised, the piling suffered rapid deterioration from the ravages of the marine borer *limnoria lignorum* and within five years it became necessary to renew the piling. This time, however, the piles were protected by scupper nails and, in recent years, some of these have been recovered. The method of protection was to drive short wrought-iron nails having large heads so closely spaced that the heads, when rusted, would provide a continuous iron layer over the wood; the length of timber to be protected would be about 10 feet and this required about 7,000 nails —probably the week's output of a woman nail maker somewhere in the Black Country!

The 7,000-ft quay wall of the New Docks at Southampton accommodates the world's largest liners. The area between the quay and the original shore line was reclaimed by pumping ashore the material dredged from the approach channel and the quay. In the foreground the liner 'Queen Elizabeth' occupies the King George V graving dock, the largest of its kind in the world.

The limitation of the Harbour Commissioners' powers under their 1803 Act and a later Act of 1810, led a group of London and local merchants to obtain an Act in 1836 incorporating a company to build docks and in 1838 the foundation stone of the Docks was laid. The site chosen by the company proved adequate for all expansion until after the First World War, being the mudland spit at the junction of the estuaries of the rivers Itchen and Test and which now represents the area occupied by the Old Docks system of Southampton. The first dock, the Outer Dock, was put into immediate use by the Royal Mail and P & O steamers when it was opened in 1842. It has been in use continuously until the present day, since its first use outgrew its 18-feet depth, by the cross-Channel steamers to the French coast and Channel Islands. In 1851 the only impounded dock in the port was opened, the Inner Dock; this was used until the Second World War for timber, grain and fruit, but that war brought destruction to most of its warehouses and since that time the dock has been used for little else but laying-up purposes. The Empress Dock, of 18½ acres water area, was opened in 1890 to accommodate the ever-growing ships which, at that time, would float at all states of the tide in its minimum depth of 26 feet. To provide the fullest use of this new dock, it became necessary, for the first time, to dredge the port approaches. The Harbour

166

Board, whose duty it was, undertook this, and from that time those approach channels have been maintained and deepened as necessary to float the ever-growing ships using the port. The present depth maintained is 37 feet below Port Low Water Datum (approximately low water springs) which allows all but the 'Queen' liners to use the channel at any state of the tide.

The Docks company, throughout its life, fought against financial difficulties, and progress was slow. Fortunately the character of the port made it possible for developments to take the form of quays for working under tidal conditions and the Itchen quays were reclaimed along the margin of the estuary of that river between 1876 and 1895. The financial struggle came to an end in 1892, when the London and South Western Railway Company acquired the Docks and, for the first time, adequate funds were available for the expansion necessary. The Test quays were constructed between 1899 and 1902 and this frontage of 2,230 feet completed the outline of the Old Docks system; behind it, in

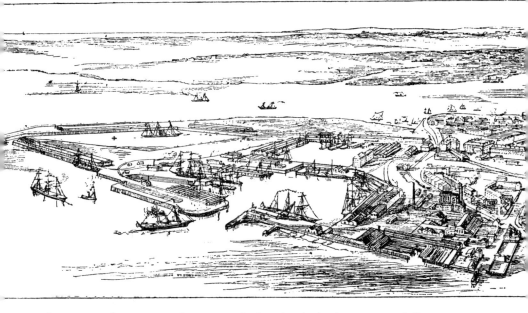

The approaches to Southampton *had to be dredged to permit full use of the Empress deep-water dock.*

A major defence port *since Roman times, Portsmouth*

Turner) has one of the longest records of continuous use.

1907, the Ocean Dock was commenced, to be completed in 1911. This dock was made necessary by the transfer of the White Star Line from Liverpool to Southampton and until 1922 it was known as the White Star Dock. Built with remarkable foresight by the engineer, F. E. Wentworth-Sheilds, it is, even today, the deepest dock in Great Britain, having a depth of 40 feet at low water springs. In spite of the development of the New Docks system between the wars, the 'Queens' still use the Ocean Dock.

The use of ships for war implies the use of ports for their building, supply and refuge, and it naturally follows that these shall be established at the most strategically convenient places. Of all the ports adopted for war purposes, Portsmouth has possible one of the longest records of continuous use. Its situation in the middle of the south coast of England, facing the potentially hostile coast of France, yet sheltered by the Isle of Wight, made it a natural choice for a fleet anchorage, while its narrow entrance gave it a great advantage in defence.

The Romans chose it as a major port for defence against Saxon pirates and usurpers of their own race, building the great fortress Portus Adurni at Portchester on the north shore of the harbour. Following the Norman invasion it became, for 138 years, with Southampton, an important means of communication with Normandy, but the loss of that connection by King John in 1204 made Portsmouth once more of strategic importance. During the reign of John, the first elementary works for docking ships were constructed, consisting probably of mud walls raised around ships after they were drawn up for beaching at high water.

Apart from its use as a place of assembly for ships—and even then it was of less value than Southampton, which had been better maintained for its commercial use—Portsmouth was not developed again until the coming of the Tudor kings, whose policy, since continuously followed, was to maintain naval strength at all times. The first dockyard works undertaken by Henry VII under this new policy were constructed at Deptford from 1485 onward, and about the same time the area of Portsmouth was heavily fortified. On 14 June 1495, Robert Brygandine started construction of a new 'dokke' for the king, which was, in the way we apply the term, the first dry dock ever made in England, and probably in the world. The payment of £193 0s. 6¾d. for its total cost is described in Brygandine's patent as follows: 'Payments upon the reparalleling, fortifying and mending the dokke for the King's ships at Portsmouth, making of the gates and fortifying the head of the same dokke.'

The work seems to have been well planned, as it was completed without accident or delay in forty-six weeks in the years 1495–6, operations being held up for two periods during that time. There was a large amount of timber used in the dock, probably for the sheet pile lining; 664 tons of stone were used to strengthen the dockhead. The royal ship *Sovereign* was docked there and to get her out, twenty men were engaged for twenty-nine days 'at every tide both day and night, weying up of the piles and shorys and digging of ye clay and other rubbish between the gates'. There were apparently two single gates separated by a space which was filled with clay puddle, as a document refers to 'ye inner as well as ye uttermost gate'. One 'ingyn' was used to empty the dock. Construction time was twenty-four weeks for the dock, twenty-two weeks for the gates and dockhead, with the number of paid men varying from sixty to twenty-eight. Labour cost approximately the same as at Southampton sixty years earlier, with carpenters receiving 4d. to 6d. a day, sawyers 4d., and labourers 3d. Iron cost from £3 14s. to £4 per ton.

During the first part of the reign of Henry VIII, Portsmouth grew in importance as a naval port and the dry dock was used to the fullest extent, being enlarged and reconstructed to take a ship of 1,000 tons—*Henry Grace à Dieu*. In 1545, the French made a landing in the Isle of Wight and their fleet lay off Portsmouth. This led to a partial eclipse of Portsmouth and the rise of Deptford, Woolwich, and Chatham so that although £73,305 was spent on Deptford between 1559 and 1570, only £6,641 was spent at Portsmouth in the same period. The concentration of naval dockyards in the Thames and Medway remained a part of naval policy until the Commonwealth, when the Thames estuary proved as vulnerable to the Dutch as Portsmouth had been to the French, while both dockyard areas were of importance in the fleet's task of denying the English Channel to the enemy of both nations. Portsmouth, however, was close to the New Forest and the Forest of Bere, but the Thames dockyards had practically no local supplies of timber. A report to this effect made by Sir Anthony Deane in 1666 and the appointment of Samuel Pepys as Secretary to the Admiralty led, from 1684, when he began his second term of office, to a rapid and continuous rise in the development of Portsmouth. By the end of the seventeenth century Portsmouth had become, as it still remains, the leading Royal Dockyard and naval port; its relative importance at the end of the eighteenth century may be assessed from the expenditure in 1783 of £65,198 on the dockyards of Portsmouth, Woolwich, Chatham, and Plymouth of which sum no less than £43,000 was spent on Portsmouth. This century saw the growth of the dockyard area from about 26 acres to 82 acres, partly by the acquisition of land on the foreshore area, but mainly by reclamation; the nineteenth century saw this grow by similar means to about 270 acres, almost its present area of nearly 300 acres.

Many famous names are associated with the development of Portsmouth Dockyard. George Villiers, Duke of Buckingham, persuaded the king, in 1627, that the harbour and dockyard should be restored, but Phineas Pett, the famous shipbuilder, recommended otherwise and nothing was done. The activities of Samuel Pepys have already been referred to, and there is little doubt that his influence, more than any other, made Portsmouth what it is today. The beginning of the nineteenth century brought four great engineers into the affairs of Portsmouth; John Rennie, who had been consulted on many works for the Admiralty about that time; and Sir Samuel Bentham, the brother of Jeremy, the economist, who developed the use of steam driven wood-working machinery and introduced it into shipbuilding first at Redbridge, near Southampton, and then at Portsmouth. Bentham was responsible for the introduction of the block-making machinery developed by Marc Isambard Brunel and built by Henry Maudslay. This amazing advance in machine tools consisted of a complete series of thirty-two machines for the making of ships blocks; installed in 1808, many of them are still in the

Portsmouth block shop and some still in use, the oldest surviving mass production factory in the world.

Essential as harbours may be to the mariner, equally important are those means of identification which enable him to find them and, at the same time, to avoid the hazards to be found on every coast. In the century or so since steam navigation came into effective use, the hazards of seagoing have been so reduced that there is a tendency to forget the risks taken by the mariner in earlier times. Between 1800 and 1880, Britain alone was losing ships at the rate of between one and three a day, and most of these were lost when nearest to safety. It can be appreciated from this how important identifiable features on the coast could be to the mariner and this has been reflected in the laws of all countries dependent on sea communication. The spires and towers of churches, in particular, have been used as means of identification around the coasts as, for instance, the twin towers of the Saxon church at Reculver, previously the Roman port of Regulbium, which were for many years preserved by the Admiralty as a

achinery for the making of ships' blocks, developed by M. I. Brunel, as installed at Portsmouth in 1808.

navigational landmark and which are now cared for by the Department of Ancient Monuments. Not far away is all that remains of the Roman *pharos* or lighthouse at Dover, successor to the Pharos of Alexandria, ancestor of all lighthouses and one of the seven wonders of the world. This building, attributed by the Roman Strabo to Sosostratus, and built in the reign of Ptolemy II, was sited on an island about a mile to the north of the port and approached by a causeway. The construction was divided into three parts: first, a base in the form of a truncated pyramid about 220 feet high; on this was an octagonal pyramid section of 100 feet; and at the top, a cylindrical section of about 30 feet carrying the lantern or brazier.

The development of navigation in the sixteenth century drew attention to the danger to ships entering the estuaries of the

A causeway *connected the Pharos of Alexandria with the mainland, as seen in this 16th-century engraving.*

174

The twin towers *of the Saxon church at Reculver (left) and the highly elaborate tower on the Rock of Cordouan afforded landmarks for navigators.*

Garonne and the Dordogne through the presence of the Rock of Cordouan. To reduce this hazard, King Henry II of France in 1584 commenced the building of a great lighthouse on the rock which was finished by Henry IV in 1610. The tower, 115 feet high and 50 feet diameter at the base, was built on a solid platform of masonry, 100 feet in diameter. The whole erection was highly elaborate and quite unlike the functional form of tower later developed for the same purpose elsewhere.

The first serious interest in Britain, other than that of the Romans, developed with the growth of sea power in Tudor times. Henry VIII, who built the chain of defensive castles around the coast, appreciated the need to carry the battle to the enemy and thus initiated Britain's great naval expansion. For the governing of matters appertaining to the sea, he created two new bodies: one a board to administer the King's Navy Royal, now the Board of Admiralty; the other a technical body of experts to advise on matters of ships, seamanship, navigation and such matters to whom, in 1512, he granted a licence 'in honour of the

Holy Trinity and St Clement in the Church of Deptford Strond for the reformation of the navy, lately much decayed by admission of young men without experience'. In 1514, the Guild was granted a charter by the king with the title of 'The Brotherhood of the Most Glorious and Undividable Trinity', with extensive powers to improve conditions of navigation. These powers were continued until the reign of Elizabeth I, when an Act of Parliament strengthened the authority of the Master, Wardens, and Assistants of Trinity House by empowering them to control and license the Thames pilots and also to establish such marks or beacons around the sea coast as may in their opinion be required, to make such charges for this, and constituting it a criminal offence to damage or remove such marks. These powers, were, in 1593, made exclusive to Trinity House, but it took nearly two and a half centuries for the monopoly to be made effective. The years dragged on with disputes and litigation between private owners of beacons and the Brethren of Trinity House while ships were wrecked in great numbers on the unmarked coast.

Such a state of affairs was bound to come to an end, and the first real advance arose as a result of the rapidly growing importance of the port of Plymouth; from the natural haven for the ships of Elizabeth I, it had, by the end of the seventeenth century, become a centre of trade and, under William of Orange, a favoured port for naval activities. One thing impaired the progress of the port—the Eddystone Rock, a reef of gneiss which, standing above the sea bed 14 miles off the shore, took its regular toll of English Channel shipping. From the mid-seventeenth century the pressure had been rising in Plymouth and, in 1664, Trinity House was requested to authorize the erection of a lighthouse on the Eddystone, but this application was refused. In 1694, Trinity House capitulated and a Patent was issued in that year for the erection, by private enterprise, of 'a Light house or Beacon with a light upon the Rock called the Eddystone off Plymouth', the cost to the owners being recouped by tolls on shipping, from which Trinity House was to get a proportion.

For a year no person could be found to undertake such a hazardous operation as the erection of a structure on the Eddy-

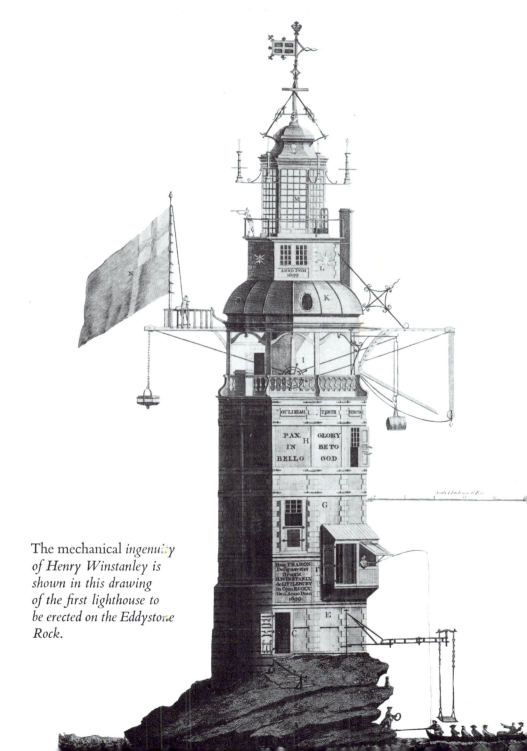

The mechanical *ingenuity* of Henry Winstanley is shown in this drawing of the first lighthouse to be erected on the Eddystone Rock.

Rudyerd's 'Edystone Light House' *was made of timber and ballasted to the rock with a stone core.*

stone rock, but in 1695 Henry Winstanley, a shipowner who had acquired a reputation for mechanical ingenuity, offered to carry out the work. He spent the winter of that year preparing his plans and in the summer of 1696, under conditions of great difficulty, succeeded in securing twelve iron anchorage bars in holes cut for them in the hard rock. The next summer saw the erection of a stone base, 14 feet diameter and 12 feet high; this was enlarged in 1698 to 16 feet diameter and 18 feet high and on it was built a polygonal tower surmounted by the lantern. On 14 November in that year Henry Winstanley went out to his tower and lit the candles, the first light ever shown on the Eddystone and the first lighted beacon ever displayed on such an exposed rock. The first winter on the rock demonstrated that the structure needed strengthening, and Winstanley enlarged the structure, raising the height to a total of 120 feet to the weather vane. The winters of 1699 to 1703 were memorable ones to mariners; not a single ship was lost on the Eddystone, but, in the five years, the tower had taken a terrible battering and on the night of 26 November 1703, during one of the greatest storms ever known on the coast of Britain, Winstanley's tower was one of the many casualties of the gale. The builder, at fifty-nine years of age, had achieved his great ambition to be on the tower in a great storm, and, with his lighthouse-keepers, was lost with his lighthouse. Two nights later the large merchant ship *Winchelsea*, on her way home from Virginia, struck the rock and was lost. Two survivors got to Plymouth with the news and this incident proved the urgency of restoring the light.

In 1706, a new tower was put in hand under the supervision of its designer, John Rudyerd, a silk merchant of Ludgate Hill, London, and an inspired amateur in engineering. Rudyerd was assisted in the execution of his scheme by the very practical knowledge of two master shipwrights, Smith and Norcutt, whose experience was essential, as Rudyerd's design consisted of a timber tower of conical shape, ballasted down on the rock with a stone core. The tower was secured to the rock by thirty-six iron straps, secured 16 inches deep in the rock with melted pewter. No effort was spared to expedite the work and the tower was completed and the light burning in the summer of 1708. For fourteen years, the lighthouse functioned without trouble. Then, in 1723, it was found that the lower ends of the timber casing had been attacked by marine borers, probably the *limnoria*. This is a small shrimp-like creature which, by multiplying rapidly and consuming the wood from the outside, eventually reduces the section; in fourteen years the lower ends of Rudyerd's timbers were probably reduced to practically nothing. The timbers were repaired and, from that time were regularly maintained, first under the supervision of a shipwright, John Holland, brought from Woolwich

Dockyard and, later, by a local shipwright, Josias Jessop. It was by this means that the lighthouse was kept in good working condition until the night of 1 December 1755, when fire destroyed the pitch-saturated timbers of the tower, bringing the masonry core down with them and Rudyerd's lighthouse, after a century of service to mariners, ceased to be.

It is fortunate that the lighthouse was owned by a body of proprietors who were prepared to leave the organization of re-building to one of their number, Robert Weston, a man of initiative who immediately sought the advice of the President of the Royal Society, the Earl of Macclesfield. The President recommended John Smeaton, who at twenty-eight years of age had been elected a Fellow. John Smeaton, the son of a lawyer, was born at Austhorpe Lodge, near Leeds, on 8 June 1724, and at an early age showed signs of strong mechanical inclinations. He went to Leeds Grammar School and, in 1742, although it was against his father's inclination, was apprenticed to a London mathematical instrument maker. He soon extended his range of interests, and, before long, was undertaking all kinds of mechanical work. His Fellowship of the Royal Society brought him into contact with men of the professional class and, in 1759, he presented to the Society his famous paper 'An Experimental Inquiry concerning the Natural Powers of Water and Wind to turn Mills and other Machines depending on a Circular Motion'. This work, based on his study of the writings of Continental engineers, on his travels in the Low Countries and, above all, his own experiments, marks the beginning in Britain of the profession of the civil engineer and, quite justifiably, brought Smeaton the Gold Medal of the Royal Society.

Smeaton accepted the invitation to design and supervise the construction of a new Eddystone tower and, against current opinion, decided to construct it entirely of stone. He modelled the shape on the trunk of an oak tree, considering that the base of such a tower should be spread over the greatest possible area while, at the same time, the shaft, like that of Rudyerd's, should offer the minimum resistance to the sea and wind. Unlike the stone base of Winstanley's tower, which relied entirely on mortar for the bond,

Built entirely of Portland stone *with granite for the outer cladding,
John Smeaton's tower was modelled on the shape of an oak-tree trunk.*

Smeaton secured his blocks together with interlocking dovetails
and iron cramps. He put his ideas in the form of a model, which
was submitted to and approved by the Lords of the Admiralty.
Smeaton then went to Plymouth to put the work in hand and
appointed Josias Jessep, the shipwright, as his resident engineer.

This appointment started a friendship which lasted until Jessop's death, after which Smeaton took charge of the education of Jessop's son, William, and trained him as a civil engineer. William Jessop became a most influential member of the profession and, not only by his own works, but by the advice which he freely gave to younger engineers of whom Telford and Rennie are examples, confirmed for the future the high professional standards inculcated in him by John Smeaton. Even today, the importance of the Jessop link between Smeaton and the professional engineer is not fully appreciated.

Smeaton decided on Portland stone for the bulk of his masonry, with granite for the outer cladding. The stones were supplied cut to approximate sizes from the quarries, to be finally cut and fitted at a work yard established by Smeaton at Millbay, the site of the present docks of the British Transport Commission. All the courses were cut to templates which had been previously fitted together on a large floor, so that assembly on the rock would go on without difficulty. On 12 June 1757, the first stone, weighing $2\frac{1}{2}$ tons, was laid on the prepared foundation and on the next day, three more stones to complete the first course were laid; by the end of the month, seventeen stones, representing the first two courses, had been laid. From then on, each course represented an improvement in the conditions of work, and, as each summer season passed, the tower grew until, on 16 October 1759, the Eddystone light was again in being. Smeaton's tower remained the Eddystone lighthouse until it was replaced in 1882 by the present tower of Sir James Douglass. It was not through any defect of Smeaton's work that a new lighthouse had become necessary, but through the undermining by the sea of the rock ledge on which the tower stood. The present tower, built on an adjoining rock, stands twice as high as Smeaton's which, when its life on the rock was finished, was taken down to the plinth level and re-erected at the expense of the citizens of Plymouth on a new base on Plymouth Hoe where it stands today, a fitting memorial to John Smeaton.

Egyptian hydraulic engineering — Roman aqueducts
monastic water supply in Britain — Drake's
leat to Plymouth — water supply for London
Middleton's 'New River' — the war against disease
irrigation and electricity generation

◄The street cries *of the water carriers were heard frequently in all the*
great cities of Europe until main water became, as it had not been
since Roman times, easily available.

Water Supply and Public Health

It is a characteristic of our modern times that many of the essentials to our wellbeing are so regularly and unpretentiously supplied that they are taken for granted. Of these, perhaps the most noteworthy is the abundance of pure cold palatable water. This commodity was considered of the greatest importance by our ancestors in all past ages. Primitive man, of necessity, went to the water, not possessing the means to take it with him. As communities developed, sooner or later the population outgrew the natural resources of the immediate locality and means had to be devised for bringing supplies from a distance. The greater population would also demand more food, and as intensive cultivation in hot countries implies some means of irrigation, hydraulic engineering developed early in the great civilizations.

In Egypt, the Nile is, and always has been, the essential source of life-preserving water. Its control has, therefore, been a major objective of the country's engineers from nearly 6,000 years or so ago, when the civil engineers of the time built a dam of hewn stone at Kosheish to divert the course of the Nile so that Mena could build Memphis on the site of his choice. The engineers of Usertesen III enabled him to sail southward to defeat the Ethiopians about 2660 B.C. and during the same period the officials appointed by the king included one with the title of 'Royal Irrigation Superintendent'. This official and his successors, or their equivalents, must have continued to function for many centuries as, by the time of Rameses II, whose reign ended about 1322 B.C., the Nile valley had acquired an extensive pattern of artificial lakes or reservoirs, with the distributing canals necessary to distribute water over a large area.

About the same period as that of the Egyptians the people of the Persian Gulf area were performing equally extensive hydraulic

Baked clay *was used for this drain pipe in the great Indus city of Mohenjo-daro, built some 4,500 years ago.*

works, as in Babylonia when, in 2350 B.C., Siniddinan, the King of Larsa, enlarged a canal, already ancient, and his neighbour Rimsin, the King of Elam, at about the same time opened up an outlet for the Tigris to the Persian Gulf. His efforts were augmented about thirty years later when Khammurabi, King of Babylon, organized an extensive system of flood control canals to ease the burden of the inhabitants of the lower Tigris valley.

In India, the conservation of water was undertaken from a very early date, while in Baluchistan the Ghorbasta, great cyclopean stone dams, were constructed about 1800 B.C. Persia, also, was served by the hydraulic engineer by the construction of irrigation works between about 700 B.C. and 600 B.C. and soon after, in about 520 B.C., Athens was supplied with water from an ancient spring, Callirhoe, which flowed from the foot of a rocky ridge crossing the bed of the Ilissus.

The Romans took their water supplies very seriously, going to great trouble and expense to bring adequate quantities of pure water to all their city dwellers. Even their camps were sited with great care; the preliminary survey included tests of the soil and water, and checks for levels so that adequate drainage would be assured and so that the defence ditches would hold adequate water without excessive excavation. Although many Roman writers make reference to water supply and the works in connection therewith, later generations are mainly indebted to a great water engineer for his very full and accurate account of the Roman achievement in his branch of the profession. He was Sextus Julius Frontinus, who was the Water Commissioner of Rome from A.D. 97 to 104 and whose Latin manuscript was preserved for many centuries in the monastery of Monte Cassino, destroyed in the Second World War. Fortunately for English readers, a photographic reproduction and English translation were published by Clemens Hershchel, an American engineer. No man could be more conscientious in the performance of his duties than was Frontinus and it would have delighted his heart if he could have known that, for many centuries, his writings, intended for the guidance of his immediate successors, would be used as the main authority for the practice of water engineering.

Rome's first aqueduct, *built by Appius Claudius in 313 B.C.*, *brought water to the city from a spring several miles away.*

Frontinus tells us that for the first 441 years from the foundation of their city, the people of Rome were supplied with water from private wells and from the river Tiber. In 313 B.C., however, the supply was augmented by the Censor, Appius Claudius Crassus who not only built the Appian Way, but brought water by Rome's first aqueduct from a spring between 7 and 8 miles from the city. The length of the channel was about 10½ miles, of which all but 300 feet was laid underground, the remainder being on a masonry or arched structure above ground, the section being 5 feet high by 2½ feet wide. This aqueduct was followed by the construction in 273 B.C. by Curius and Fulvius of the Anio Vetus, the 'Old Anio', over 40 miles long and served by a catchment area of about 150 square miles. These two aqueducts were adequate for a century, but by 145 B.C. they were both losing water through leakage and by theft, or unlawful extraction. The Senate therefore commissioned the Praetor Marcius Rex to restore them and to investigate other possible sources. He recommended bringing water from a source near to the thirty-eighth milestone on the Sublasclensian Way 'where numberless springs gush forth from caves in the rocks immovable like unto a pool, and of a deep green hue'. There was, after reference to the Sybilline books, some objection to the use of this source, but three years later the Senate authorized the construction of the Marcian Aqueduct. It was 58½ miles long, of which 51⅓ were underground and contoured the hillsides, half a mile was in a masonry channel above ground and 6½ miles on arches; the section was 3½ feet wide by 5½ feet high. The water from the Marcian springs was restored to Rome nearly a century ago by a modern aqueduct called 'Acqua Pia'.

The fourth aqueduct, the Tepula, was constructed in 127 B.C., its name being derived from the tepid character of the spring water which, at 63° F., in more recent times acquired a reputation for possessing curative properties. The Tepula was almost entirely of concrete construction, the earlier ones having been built of traditional masonry, and its completion so improved the Roman water supplies that no further additions were needed for ninety-three years. By this time, 33 B.C., the Appia, Anio and Marcia aqueducts were in a ruinous condition; Agrippa therefore rebuilt them and added a fifth aqueduct, the Julia, to connect with the Tepula. Thirteen years later he made another addition to the city's water resources by constructing the Virgo, so called because a young girl pointed out the springs to a military survey party who were prospecting the district.

The seventh aqueduct was adversely criticized by Frontinus; known as the Alsetina, it was constructed by the emperor Augustus to provide water for a 'Naumachia' (a naval equivalent of the Aldershot Tattoo) and for irrigation. The poor quality of the water brought in, quite unsuitable for domestic purposes, appeared to Frontinus to be a waste of the engineer's efforts. The increasingly high standard of cleanliness in Rome, amounting in some cases to luxury, led in due course to a further demand for water and, in A.D. 36, the emperor Tiberius began two more aqueducts which were completed by Claudius in A.D. 50, to be named by him Claudia and Anio Novus. All these nine aqueducts were used for the supply of Rome, and are described by Frontinus. There was one other, the Aqua Crabra, which did not reach the city and was not included in Frontinus' description, although it existed in his time. After his days, another four aqueducts were added to the nine; they were the Aquae Trajana, Alexandrina, Septemiana and Algentia.

With the exception of the Virgo, Appia and Alsetina, the aqueducts were provided with settling tanks (*piscinae*) in which sediment would settle. In the city, large reservoirs called *castella* held a reserve for distribution to smaller *castellae* which, in their turn, supplied the various users. Delivery to the user was regulated by the size of the aperture through which the water discharged, no allowance being made for velocity due to the head. The standard unit was a *quinaria* or circular outlet equal in area to 0·632 of an English square inch.

The system of aqueducts was destroyed by the Goths when, in the fifth century A.D., they besieged the city and Rome was left without adequate water resources until after a partial restoration by Pope Adrian I and his successors from A.D. 776 onward. In the early 1870s, a British company undertook to run water to the city through underground mains from some of the early sources and, ten years later, water from the sources of the Marcian aqueduct was restored by that means. From that source alone, it is estimated that the upper half of the original Marcian aqueduct carries 27 million gallons a day which, from a point above Tivoli, is carried underground by several lines of 24-inch cast-iron pipes. Three of the old aqueducts still provide Rome with a part of its water, the Acqua Vergine (Virgo), Acqua Felice, part of Alexandrina which was restored by Pope Sixtus V in the sixteenth century, and Acqua Paola, restored by Pope Paul V in 1611.

Water supplies were important to the many communities set up in other parts of the Empire; the monumental relics left in parts of Europe, the Middle East, and Africa bear testimony to this. It is estimated that about 200 examples of Roman aqueducts are still to be found, either complete or, more often, in ruins.

The aqueduct *at Segovia, known to the Spanish as the 'Devil's Bridge', probably represents the peak of Roman engineering achievement. The 3-tiered Pont du Gard (right) near Nîmes in southern France, was built about 18 B.C. The arches are laid entirely without cement.*

Such relics include the Pont du Gard, near Nîmes in Southern France, part of a 25½-mile channel, mostly in tunnel, constructed by Agrippa about 18 B.C. The bridge carrying the aqueduct is formed of three tiers of arches crossing the River Gard at a height of 160 feet and is 885 feet in length. The arches, varying in the two lower tiers from 51 to 80¾ feet and uniformly 11½ feet in the top tier, are laid without cement; the channel, however, was lined with that material and is 4 feet wide and nearly 5 feet high. The aqueduct at Segovia, built a little more than a century later,

probably represents the peak of Roman achievement in engineering structures. It is approximately 2,700 feet long and the highest arches are 119 feet high. There are two tiers of arches, each about 15 feet span, the bases being only 8 feet wide. Known to the Spanish as the 'Devil's Bridge', the aqueduct was restored in the fifteenth century by Queen Isabella to bring water from the River Frio in the Sierra Guadamarra, 10 miles away.

After the time of the Romans, the habit of cleanliness lapsed and, as their works deteriorated or were destroyed by the bar-

barians, populations reverted to original sources for their supplies, polluted though they might be. A few monastic communities, and they were rare, maintained supplies of fresh water for their own benefit and, in some cases, that of the surrounding community. The earliest known example in Britain of monastic water supply is that of the Benedictine priory of Christ Church, Canterbury (the cathedral). The work was carried out by Prior Wilbert in 1160 and the original plans are still in existence. From the evidence of these plans, and of the general engineering practice of ecclesiastical water engineers of mediaeval times, it appears that the writings and work of Frontinus and his contemporaries was held in great respect by learned men of the Church.

A notable example of a monastic water supply is that of the Friars Minor (the Franciscans) of Southampton who, in the year 1290, were granted a licence to enclose with a stone wall the fountain of Colwell in the manor of Shirley. The spring head, enclosed in three vaulted chambers, was connected by a large diameter lead pipe to the water house, still standing in Commercial Road, where other supplies were later collected. From there it was carried in lead to the friary itself, a distance of about two miles from the conduit head, and divided between the monastery itself and God's House, a mediaeval hospital. Twenty years later, the Friars gave the surplus water to the town 'out of reverence to Henry, Archdeacon of Dorset, and their goodwill to the community of the town'. By the fifteenth century, the Friars were no longer able to afford the upkeep of the water supply and, in 1420, it was conveyed to the town, from which date the burgesses of Southampton have accepted the responsibility of supplying their town with water.

Although the monasteries had supplied many communities with water and, possibly, others with the knowledge to enable them to increase their own supplies, many cities and towns suffered from a lack of the essential commodity. The town of Plymouth had, for centuries, relied on wells for the main source of supply and when in the time of Elizabeth I the port had become of strategic importance, Sir Francis Drake supported a

The method of
*boring wooden water
pipes and pump
barrels is shown in
this 16th-century
woodcut.*

The cast joint *in this
lead water pipe is
probably medieval
but is similar to
lead pipes used in
Roman times.*

Wooden water *pipes
were used extensively
in the 18th and early
19th centuries. These
sections of elm were
excavated in South-
ampton.*

193

Sir Francis Drake *in 1589 contracted with the Corporation of Plymouth to bring water to the city from the River Meavy on Dartmoor. He received for his trouble £300 and the right to establish six mills on the course of the new leat.*

scheme promoted by the municipality for bringing water in from Dartmoor. In 1585, the twenty-seventh year of the Queen's reign, an Act was passed to enable the city to construct a leat or channel to carry water into the town. The purposes were, according to the Act, for watering ships and protection against fire; also to scour out the harbour and remove refuse from the tinworks and mines. The proposal was to build a weir on Dartmoor across the River Meavy and to divert the water into a ditch or trench 6 or 7 feet wide, but of no specified depth. The direct distance of the weir from the town was about 10½ miles, but in following the contours, the length of the channel was nearly 18½ miles.

Drake had been Mayor of Plymouth in 1581, but did not become interested in the proposal until 1585, when, as Member of Parliament for Bude, in Cornwall, he was a member of the Select Committee which approved the Bill. His activities for the next few years were more concerned with the sea and the Spaniards than with fresh water; 1587 saw him at Cadiz, to the annoyance of the King of Spain; 1588 found him, with the other seafarers of England, in the English Channel settling the Armada problem, and not until 1589 was he free to engage himself in business of a private nature. In that year he contracted with the Corporation of Plymouth to bring the Meavy water to the town, and to receive for his trouble the sums of £200 for the construction, £100 for land compensation and the right, tenable for sixty-seven years, to establish six mills on the course of the new leat. This concession was bitterly opposed by the owners of existing mills on the Meavy who, in 1592–3, appealed to Parliament against the Plymouth Water Act. The protest was rejected after review by a Select Committee of the House, the chairman of which, by a strange coincidence, was Sir Francis Drake! The leat proved to be of great advantage to Plymouth and, with minor modification, continued to be the town's main water supply for more than 300 years until, in 1898, the Burrator reservoir scheme was completed.

The growth of population in big cities constitutes a regularly recurring problem of catching up with an inadequate water supply. The citizens of ancient Rome increased the supply at intervals over a period of more than 500 years, London has been importing water from outside its populated area for 750 years, and, in more recent periods, the expansion of the industrial areas of the Midlands and North of England have led to their water authorities bringing supplies from the mountain areas of North Wales and Lakeland.

The growth of London's population led to the citizens drawing water from Tyburn in 1237, Paddington in 1440, Hackney in 1559 and Hampstead in 1589. These sources were exploited because the growing population, creating by its expansion a greater demand, also covered much of the catchment area of the city with dwellings, thus polluting the water by cemeteries and cesspits in the gravel. Conduits from outside to some degree overcame these difficulties, but final distribution was still by the human water bearer and in times of drought he had little to carry.

The sixteenth century brought the beginnings of a mechanical age and with it the idea of pumping water was effectively revived. The Romans had used barrel pumps as well as the Archimedean screw; a bronze pump of their period is in the British Museum, and a wooden example with two cylinders was found at Silchester. There is an entry in the records of the City of London dated 1406 referring to the 'Ordinances of the Mistery of Forcemakers' from which it appears that the 'forcer' or pump makers had attempted to form a guild. Agricola shows that, by this time, the foot valve had been added to cylinder pumps, thus making them effective for suction as well as forcing, but not until Evangelista Torricelli explained in 1643 that the atmosphere had a part in the suction process was there an understanding of the limiting suction head of about 28 feet.

Although several attempts were made about the end of the sixteenth century to pump water from the Thames by machines operated by horse gins, none of these survived for very long, and the first scheme to have any lasting success was that of Peter Morice, or Morris, described by Stow the historian as a Dutchman or a German, both words meaning much the same in those days. It is even possible that Morice was a mining engineer of British origin, who had gained experience of pumping while working in Germany where, it appears, bridge waterworks originated at Danzig in 1570. By the year 1578, Morice was in the service of Sir Christopher Hatton, Captain of the Queen's Guard and later Lord Chancellor of England, who used his influence to obtain for his servant a patent which, dated 24 January 1578, states that 'Peter Moris hath by his great labor and charge found out and learned the skill and coning to make some new kynde and manner of engynes to draw up waters higher than nature yt selfe' This patent was for twenty-one years and was conditional on the building of an engine within three years of the date of the grant.

Morice built his pump as promised and demonstrated its efficiency by forcing a jet of water over the steeple of St Magnus' Church, London Bridge. This induced the City Fathers to agree to finance him and, although he had some difficulty in keeping them to their promise, they eventually, on 30 May 1581, granted him a lease of the first of the arches of London Bridge at the city

The method of *spigot and socket jointing in water pipes was known to the Romans.*

end for a period of 500 years at a rent of ten shillings per annum. The construction was completed by December 1582, and on Christmas Eve that year London had its first regular supply of pumped water from the Thames, and the satisfaction of the city Corporation was expressed by the granting to Morice of a lease for the same period of the second arch.

No contemporary plans or description have been found of the first machine and it was not until 1635, when John Bate a well-known mechanician had an opportunity to view the machinery, that any details were made public. The sketch made by Bate suggests that the water wheel was supported by a frame through which the shaft of the wheel passed. The end of the shaft carried a crank which, through a connecting rod, gave semi-rotary motion to a disc which was coupled by pitch chains to a half disc above. These pitch chains provided a correct rectilinear motion and were coupled to the piston rods of the pumps which, according to Bate's sketch, were two in number, but might easily have been more. Some remains of pump machinery were recovered from the bed of the Thames when the excavations for the foundations of New London Bridge were being carried out in 1828 and these included a pair of pump barrels $5\frac{3}{4}$ inches diameter by about 4 feet long, together with a large square iron shaft and cross arm.

With the exception of a stoppage of about two or three years following the destruction of the plant in the Great Fire of London, the Bridge waterworks maintained a supply to its customers, but, by the end of the seventeenth century the machinery had fallen into bad repair and the descendants of Morice sold their rights for the sum of £38,000 to Richard Soane who formed a new company with a capital of £150,000. The new company, in about 1702, engaged George Sorocold to be their engineer for the reconstruction of the pumping machinery and, in doing so, employed one of the great water engineers of all time. Little is known of the life of this man, except that in all his works he put perfection first and profit as a secondary consideration. It cannot be wondered, therefore, that he was responsible for a number of waterworks, all similar to that of London. These included the following: Derby, 1692; Leeds, 1694; Norwich, 1694; Exeter, 1695; Bristol, 1696; Sheffield, 1697; Bridgnorth, 1706.

In 1767, the second arch at the south end of the bridge was provided with pumping plant by John Smeaton, whose design, although based on the same principles as the machines of Morice and Sorocold, was typical of the soundness of design and construction associated with the greatest engineer of his day. The old London Bridge was dismantled when the new Rennie bridge was built, and, to settle the affairs of the London Bridge waterworks, an Act of Parliament was passed in 1822 enabling the proprietors to transfer their rights to the New River Company for the annual sum of £3,750, or £2 10s. per share paid until the year 2,082 or 500 years after Peter Morice's first concession. The payment still goes on and no doubt will do so for the remaining 120 years.

The London Bridge source served an area limited by the capacity of the pumps and the cost of the network of mains. By the beginning of the seventeenth century, however, London had already begun its expansion beyond the city boundaries; the built-up areas of the city and Westminster had joined up and growth into the neighbouring countryside was beginning. The

'Sir Hugh Myddleton's Glory'—*the inauguration in September 1613 of the 'New River' at Islington.*

Mayor and Corporation of the City of London in 1606 obtained powers to bring fresh water from the springs of Chadwell and Amwell in Hertfordshire, but hesitated to invest the very substantial sum of money which the scheme was expected to cost. It was left to a public-spirited citizen, Sir Hugh Myddleton, a goldsmith, to take the initiative and undertake the scheme as a private venture with twenty-eight partners or 'Adventurers' to share the cost. Myddleton promised to complete the work within four years from the date of his offer, 28 March 1609; he might have done so but for fierce opposition by the landowners and the usual financial difficulties arising from an over-optimistic estimate of the cost. His work had fortunately aroused the Royal interest and, on his application, the king paid half the cost on condition of receiving half the profits. This enabled the project to be completed and, on 29 September 1613, the inauguration was celebrated in great style.

George Sorocold's *tide-driven pump under London Bridge was designed on the same principles as Peter Morice's water wheel which it replaced.*

The 'New River' as Myddleton's leat became known, was an open channel, 10 feet wide and averaging 4 feet in depth. It followed the contour line from the Chadwell Spring to a circular pond at Islington, known as the New River Head, a channel length of 38¾ miles between points 20 miles apart, but later improvements have shortened the channel to 24 miles. The fall was 2 inches to a mile and, to maintain an even gradient, it was necessary to carry the water over roads and low ground by timber troughs or flushes lined with lead. The distribution was almost entirely by elm wood pipes having spigot and socket joints, although the final services were in lead. The New River Company maintained its services and its identity until it was absorbed in the Metropolitan Water Board in 1904.

The continuing growth of London's population led to the formation of one company after another for the purpose of supplying the need for water. The Hampstead Water Company, incorporated as early as 1692, ran water by gravity to the north-west areas and was absorbed by the New River Company in 1856. The Thames was the source of supply for most of the other companies and in 1675 Ralph Bucknall and Ralph Wayne obtained a patent for the right to erect a waterworks in the grounds of York House for the purpose of supplying the St James's and Piccadilly areas. The first pumps were probably horse driven, but in about the year 1713, a steam pump built by Thomas Savery, F.R.S., was installed. Savery had patented this engine in 1698 and was manufacturing it in a workshop off Fleet Street, one of the first engineering workshops in the world. Savery's engine was a hand operated version of the modern Pulsometer steam pump; its failure to become a commercial success was entirely due to the inability of technology at that time to make a vessel strong enough to take the pressure required, about 100 lb. per square inch. It was soon abandoned at York Buildings and, in 1725, an engine by Thomas Newcomen was installed and this continued working until 1731. A second, and improved atmospheric engine was installed in 1752.

The steam pump engine *designed by Thomas Savery is here seen working in a mine.*

Waterworks were erected *at the end of the 16th century in the grounds of York House, London, to supply water for the St James's and Piccadilly areas.*

The coming of Watt's separate condensing steam engine led to the formation of a number of companies to take water from the Thames and the head of water available from the more efficient pumping machinery led not only to greater areas being supplied, but also made it possible to supply the upper floors of houses.

This enabled the wealthy to instal cisterns in the roofs and provide water closets on the upper floors of their mansions. Water closets, of a kind, had been known since Tudor times or even earlier, but Joseph Bramah, the inventor of the hydraulic press, improved the system to be substantially as we know it today. The great improvements in cast iron founding made it economically possible to use that material for water mains and, although some water engineers objected to their use at first, cast-iron water mains had become standard practice by the middle of the nineteenth century.

As the number of water supply companies increased, and with them the number of intakes from the Thames, the river was becoming more and more polluted by the human waste of the vast population of the built-up areas. In 1827, a pamphlet stated that 7,000 families in Westminster and its suburbs were supplied with water 'in a state offensive to the sight, Disgusting to the imagination, and Destructive to the Health'. Many of the water intakes were almost adjoining sewer outfalls, that of the Grand Junction Water Company being separated from a large outfall by only 3 yards. Such conditions were bound to result in frequent epidemics of water-borne diseases. Cholera, in particular, became a scourge, and the more enlightened public demanded an improvement. It is not possible to assess with any accuracy the deaths caused during this period through the water-borne diseases as our present ideas of infection rest on the work of Pasteur in bacteriology which led, in 1880, to the discovery by Eberth of the typhoid organism and of the *cholera vibrio* by Koch in 1884. Before these discoveries it was believed that contagion was conveyed by inhalation of a polluted atmosphere. In 1827 a Royal Commission investigated the sanitary conditions of the Thames particularly in relation to the waterworks and, in 1828, recommended that all intakes for water supplies should be moved upstream and also that suspended matter should be removed. These recommendations were based on aesthetic and not health reasons, but, as the recommendations were adopted, the public health benefited. In 1829, following a series of experiments, James Simpson, the engineer of the Chelsea Water Company, put his first slow sand filter to work and, by doing so, started a progressive movement towards public health which led ultimately to the virtual elimination of water-borne disease in civilized countries. Simpson's innovation of the sand filter bed was but one of a series of similar contributions which led to the improvement of public health.

An important contribution to sanitary progress was made by a medical practitioner, Dr John Snow, who, in the face of ridicule and unbelief, proved by inference the relationship between

Pipes from the Serpentine *were laid for the reservoir in London's Hyde Park in 1860.*

cholera and polluted water. His investigations were associated with a series of cholera outbreaks of which one, in 1831, cost 50,000 lives with further outbreaks in 1848, 1853 and 1865 adding greatly to that number. Snow's most convincing success was in connection with an outbreak of cholera in the Soho district of London, which he proved was confined to people using water from a pump in Broad Street which was polluted with sewage from an adjoining cesspool infected by a cholera victim. He also proved similarly that users of water supplied by the adjacent supply intakes of the Lambeth and Chelsea Water Companies were equally liable to cholera until the Lambeth Company moved its intakes up the river to Thames Ditton above the highest point of sewerage discharge. Incidence of cholera among that company's consumers immediately fell to about 7 per cent of that of the users of Chelsea Company's water.

Construction of dams has provided water for irrigation as well as domestic and industrial uses. The Kariba Dam, 420 ft high, on the Zambesi River retains the Kariba Reservoir, the largest man-made lake in the world.

The knowledge gained by such investigations had still to be used to convince the country's legislators and although many devoted workers assisted towards that end, the greatest and most persistent pressure came from an administrator, Edwin Chadwick, who, after a thorough personal enquiry into existing conditions, presented a report in 1842 on 'The Sanitary Conditions of the Labouring Classes'. This report, a classic of its kind, laid down the requirements of public health as being a good water supply; the carrying away, below ground, of all human waste; and the prompt removal of all refuse from habitations and streets. He recommended that all new works in connection with the public health services should be designed and constructed under the supervision of properly qualified civil engineers An ardent supporter of the principles laid down by Chadwick was Florence Nightingale who, in one of her own reports, said: 'The true key

to sanitary progress in cities is water supply and sewerage.' Great progress was made both by legislation and by practical improvements during the second half of the nineteenth century, but old ideas died hard. Even in 1893, an eminent engineer wrote 'As British soldiers are *costly*, it ought to be self-evident that barracks and hospitals which will preserve the men in health should be provided.'

Before the end of the nineteenth century, other great cities were finding the need for water from outside sources. In 1881, Liverpool started to exploit the resources of Wales by enclosing the Vyrnwy valley, the bed of a prehistoric lake, to inundate an entire village and eventually to carry the water through 42-inch pipes a distance of 68 miles for its citizens. Manchester, four years later, carried water from Thirlmere in the Lake District, through 95 miles of tunnel, concrete conduit and pipelines; while Birmingham in 1893 commenced a water scheme which, after eleven years of work, brought the city a supply from a chain of reservoirs 74 miles away in Wales.

In the arid areas of the world, the construction of dams has enabled the civil engineer to provide water for irrigation as well as domestic and industrial uses, while the application of hydraulic engineering to electricity generation has transformed many otherwise unproductive areas into new centres of industry.

VI

Xerxes crosses the Hellespont — Caesar's bridge across the Rhine — Le pont d'Avignon — London Bridge — an early suspension bridge — bridges of the Italian Renaissance — Perronet's bridge at Neuilly — Edwards' bridges at Pontypridd the first cast-iron bridge — Telford's works at Menai and Conway — Brunel's experiments — the Forth railway bridge — development of reinforced concrete

◄Brunel's last great work, *the Royal Albert Bridge at Saltash,* carries the railway across the Tamar from Plymouth into Cornwall. On the right is the suspension road bridge, opened to traffic in 1962.

Bridges

ALL BRANCHES OF civil engineering in varying degrees find enthusiasts among people outside the profession; some, such as railways and canals, have so many followers that innumerable societies and clubs exist in their interests. No creations of the civil engineer, however, have such general appeal as bridges; volumes of folk-lore have been written on them and pictures galore painted of them. A president of the Royal Academy is reported to have said: 'People always buy pictures with arches in them: they like to look through an arch,' and John Betjeman, in an essay criticizing the bridges over the M1 motorway, expresses the view that it is difficult to make a bridge look ugly, although it was achieved in the last century by the iron railway bridge over the Thames at Charing Cross.

Corbelled arches were used as early as 3000 B.C. by the ancient Egyptians and at the same period or a little later by civilizations as widely separated as those of the Indus valley and Syria, while by about 2000 B.C. true arches were turned in brickwork for vaults and their entrances at Ur of the Chaldees. No record exists of the use of bridges to cross rivers until the ninth century B.C. when, by the evidence of an embossed gold band now in the British Museum, the Assyrians used pontoon bridges to cross rivers. Herodotus, the Greek historian, who lived between about 484 to 425 B.C. describes how the Persians, in about 700 B.C., crossed the Euphrates at Babylon, but his story expresses doubt whether this was done by a bridge or by reducing the river depth by cutting a relief channel. Herodotus was more confident in his account of the crossing of the Hellespont by Xerxes in 480 B.C. Herodotus' own description as translated by Aubrey de Selincourt (Penguin Classics) reads as follows:

'Between Sestos and Madytus in the Chersonese there is a rocky headland running out into the water opposite Abydos. . . . This headland was the point to which Xerxes' engineers carried their two bridges from Abydos—a distance of seven furlongs. One was constructed by the Phoenicians using flax cables, the other by the Egyptians using papyrus cables. The work was successfully completed, but a subsequent storm of great violence smashed it up and carried everything away.'

Xerxes beheaded the engineers who built these first two bridges and appointed others, whose work was thus described by Herodotus:

'Galleys and triremes were lashed together to support the bridges—360 vessels for the one on the Black Sea side, and 314 for the other. They were moored head-on to the current—and consequently at right angles to the actual bridges they supported—in order to lessen the strain on the cables. Specially heavy anchors were laid out both upstream and downstream— those to the eastward to hold the vessels against winds blowing down the straits from the direction of the Black Sea, those on the other side, to the westward and towards the Aegean, to take the strain when it blew from the west and south. Gaps were left in three places to allow any boats that might wish to do so to pass in or out of the Black Sea.

'Once the vessels were in position, the cables were hauled taut by wooden winches ashore. This time the two sorts of cable were not used separately for each bridge, but both bridges had two flax cables and four papyrus ones. The flax and papyrus cables were of the same thickness and quality, but the flax was the heavier—half a fathom of it weighed 114 lb. The next operation was to cut planks equal in length to the width of the floats, lay them edge to edge over the taut cables and then bind them together on their upper surface. That done, brushwood was put on top and spread evenly, with a layer of soil, trodden hard, over all. Finally a paling was constructed along each side, high enough to prevent horses and mules from seeing over and taking fright at the water.'

The Romans built their first bridges of timber, and this form of construction was retained by them for most of their bridges in occupied countries. The earliest recorded Roman bridge was built about the sixth century B.C.—the Pons Sublicus, the Bridge of Piles—famous as being the bridge held by Horatius and his two comrades against the whole Etruscan army, while behind them the bridge was being destroyed. By the second century B.C. the Roman bridges developed into stone arch structures, each arch built as a structural entity on piers solid enough to resist the side thrusts. The piers were not always solid but in some cases were provided with flood relief arches or culverts, as in the case of the Pons Mulvius which, now as the Ponte Milvio, still carries heavy traffic across the Tiber, including the tanks of the opposing forces during the Second World War. This remarkable bridge, although partly restored in brickwork during the fifteenth century, still

Rome's oldest bridge, *still used today, is the Ponte Fabricio, built in 62 B.C. to link the left bank of the Tiber with the Isola Tiburnia.*

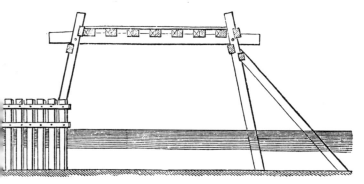

Caesar's own description *of his wooden trestle bridge over the Rhine served as the basis for this reconstruction by Palladio.*

retains two arches substantially in the original form and materials of the Romans. No doubt the Roman army, dependent as it was on its engineers, took great pride in their exploits, and this, no doubt, led to some of the almost legendary stories of their greatest achievements. Of these, the construction by Caesar's engineers in 55 B.C. of a wooden trestle bridge over the Rhine in ten days is probably the best known. According to Caesar's own description, the bridge construction allowed for the effect of fast currents carrying floating timber and the bridge trestles were suitably protected with extra piles upstream of the trestles. Andrea Palladio, the Italian architect, in his treatise on architecture, gives a convincing illustration of its construction. Another Roman bridge of which we have even more authentic information, is that built by the emperor Trajan across the Danube in A.D. 104. This is depicted among the bas-reliefs recording the emperor's life on the memorial column still standing in Rome and shows the bridge of timber truss construction on stone piers, which may be assumed to be reasonably correct as both bridge and column were the work of the same man, Apollodorus, a Greek of Damascus.

After the Romans, the art of bridge building lapsed until the twelfth century when, under the influence of the Church, bridge building revived and, in 1177 the famous bridge at Avignon, the Pont St Bénézet, was begun; part of it still stands today. St Bénézet, according to tradition, was a shepherd-boy named Benoit who was shown, in a vision, how to build the bridge and

who, to convince the sceptics of his divine inspiration, laid un-
aided a huge stone in position for its foundation. It is more probable that Benoit, educated by the Church, showed a flair for
construction and was encouraged by his brethren to supervise the
building of a much needed bridge to aid their communications
with other religious houses. From that time onward, the Church
took a great part in the building of bridges, many of which
included chapels where the faithful could give thanks for a safe
crossing, pray for the donors of the bridge and contribute towards
its future maintenance. So important did the Church consider
bridges as part of its work for the community, that its head
adopted Pontifex Maximus as one of his titles.

There is a legend that St Bénézet founded an order named the
Fratres Pontifices or Brothers of the Bridge, who were dedicated
to the construction of bridges wherever they were needed to
serve the religious and lay community. If such a fraternity existed, it may have included Peter of Colechurch, curate of a little

All that remains today *of the famous Pont St Bénézet at Avignon,
begun in 1177.*

Old London Bridge *with its street of shops and houses as it appeared before the Great Fire. The bridge itself survived until the 19th century when it was pulled down (right) and replaced by Rennie's bridge which still stands.*

chapel in the City of London, who, in the year 1176, raised subscriptions for the building of a new bridge over the Thames. This bridge, completed in 1209 after its founder's death, was the first stone bridge to cross the Thames for the convenience of the people of London. Its length was 926 feet, its breadth 26 feet carried on twenty arches of irregular size and shape, supported on nineteen starlings, or artificial islands. These starlings so restricted the waterway that during the rise and fall of the tide, the river rushed through in the form of rapids, constituting a great and costly hazard to life. The bridge carried a chapel and a large number of houses, the latter of such construction as to constitute a considerable fire risk so that they were all destroyed by fire in 1212, again in 1633, and once more in 1666, during the Great Fire of London. The houses were at last pulled down in 1754 and the bridge survived until the new London Bridge of John Rennie had been built alongside it, when it was demolished in 1832 after a useful life of 623 years in the service of the people of London.

This gateway *on the bridge over the River Monnow at Monmouth is typical of the fortified bridges of the Middle Ages.*

Rivers were often the boundary between rival states and, in any case, constituted natural defence lines against enemy forces, so that many bridges built during the Middle Ages were heavily fortified, having one or more strong gateways to impede the passage of an invader.

The new interest in bridges led to some fresh ideas, some of which may have arisen from the communications with the East opened up via the Crusades. About this time the cantilever form of construction became known in Europe although it had been used in a primitive form in the Far East for some centuries. Interest in trusses was also revived, possibly as a result of the mathematical interest deriving from the Arabic world and also filtered back through the influence of the Crusaders. Such notions were discussed, drawn, and developed by the artist-engineers of the Renaissance; men such as Leonardo da Vinci and, in particular, Palladio, saw great possibilities in the truss construction. The suspension bridge, which had been used in China since before A.D. 400, also caught the attention of these enthusaists of the new age, and the *Machinae Novae* of Fausto Veranzio, published in 1595, illustrates two examples.

The artist-engineers of the Renaissance infused into bridge design and construction that new life which was characteristic of their time. The Italian states, where the movement was born, still possess splendid examples of their achievements, although

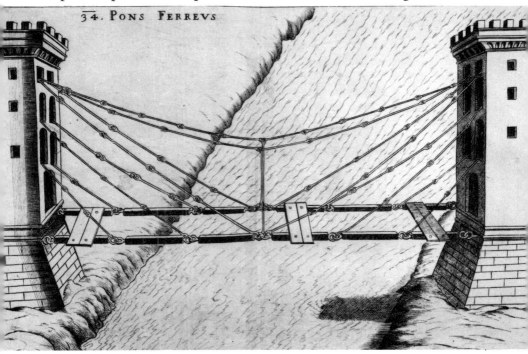

34. PONS FERREVS

A remarkable design *for an iron suspension bridge, published in 1595.*   219 nature and war have removed many others. Typical of the new approach to building, the bridges of the Renaissance break away from the traditional semicircular arch of the Romans, which, of necessity, required heavy piers at close intervals in the river bed, obstructing the flow and often, by increasing the current to a scouring rate, causing the undermining of their own foundations. The earlier Renaissance builders adopted the segmental arch, which immediately gave them the advantage of increasing the ratio of opening to pier by a factor of two. A famous example was the Ponte Castelvecchio, or Scaligero Bridge, with a span of 160 feet, built in 1354–6 over the Adige River at Verona, probably by Jean de Farre and Jacques de Gozzo. It had solid spandrels and curiously forked battlements. This bridge may have inspired the constructors of a bridge over the River Adda at Trezzo, built for Bernabò Visconti, Duke of Milan, in the years 1370–7. This was a fortified bridge providing access to the Duke's castle and crossed the river with a single span of 236 feet and nearly 70 feet rise. It survived only half a century, as it was deliberately destroyed during a siege of the castle in 1416; the abutments and some remains of the arch survive to the present day as relics of the greatest bridging achievement between Roman and modern times. The city of Florence possesses a superb example of a Renaissance segmental arch bridge in the Ponte Vecchio, the

Two celebrated Italian bridges *of the Renaissance were the Ponte Scaligero (below left) at Verona, rebuilt largely with original materials after its destruction in the last war, and the Ponte Vecchio (above) in Florence, built in 1345 for the goldsmiths.*

Bridge of the Goldsmiths. Built on the site of an earlier bridge, it has three arches varying from 95 feet 5 inches to 85 feet span, with the rise between 14 feet 7 inches and 12 feet 10 inches, the springing being about $11\frac{1}{2}$ feet above the level of low water, or approximately at high water level. The bridge is 105 feet wide, space being provided for goldsmiths' shops to be erected on both sides of the roadway, no doubt as a consideration for the provision of a substantial part of the cost of the bridge.

The scientific revival of the Renaissance led naturally to engineers becoming interested in the possibilities of other shapes for bridges including applications of mathematical curves and, between 1566 and 1569, Florence was enriched by the construction of the Ponte Santa Trinita to the designs of Bartolomeo Ammanati Battiferri da Settignano, a native of the city, who lived there, with the exception of periods of study in Venice and Rome, all his life from 1511 to 1592. Ammanati, as he became known, was appointed engineer by the Grand Duke Cosimo I, and many of the details of the construction of his bridge were preserved in a notebook kept by the contractors, Alfonso and Guilio Parigi, father and son. The foundations were constructed in the dry, behind coffer-dams which enabled the excavation to

The Ponte San Trinita, Florence, *is regarded as one of the most beautiful bridges in the world. It was built by Ammanati between 1566 and 1569. The*

arches appear elliptical but are in fact portions of two parabolic arches whose angle is hidden by the escutcheon.

be carried to the phenomenal depth of 14 feet below water level (probably low water). The foundations were laid on rows of piles driven a further 14 feet into the river-bed. The centering, of trussed construction, represents a great advance and may be a result of the cross-fertilization of ideas which must have been going on in Italy at that time. The arches appear to be elliptical but are, in fact, portions of two parabolic arches whose angle at the apex is masked by the escutcheon. The spans vary from 96 feet to 86, the rise is approximately one-seventh of the span and the width of the bridge is 33 feet 9 inches. The elevations are highly decorated with mouldings (some say over-decorated). This is a matter of taste, but few will deny that the Bridge of the Trinity is one of the world's most beautiful bridges. It was ruthlessly and needlessly destroyed by the retreating German army in the Second World War, but the inhabitants of Florence, proud of their treasure, salved the stones from the river-bed and painstakingly restored it to its former glory.

The influence of the Italian engineer-architects of the Renaissance passed in some degree to France, and was to be seen in the construction of some notable bridges of the sixteenth and seventeenth centuries, such as the Pont Neuf in Paris, one of the same name in Toulouse, the Henry IV bridge at Chatellerault, and the Pont Royal in Paris, these four bridges covering the period from 1542 to 1689. The real advance in France came during the reign of Louis XIV, a monarch of absolute power, whose expenditure matched his authority. The king's ministers, in order to maintain the royal finances, took drastic steps to encourage the growth of industry and the prosperity of the State. For this purpose, the administration became highly organized and communications to the provinces essential. To this end, the Corps du Génie was formed in 1672 and the Corps des Ponts et Chaussées in 1716; these, with the scientific advances communicated through the memoirs of the French Academy, were to advance civil engineering in France to a level, at that time, far in advance of any previously known. Bridge engineering, in particular, was put on a sounder basis by the writings of such men as Hubert Gautier, who, in 1716, published his *Traité des Ponts* which became the standard

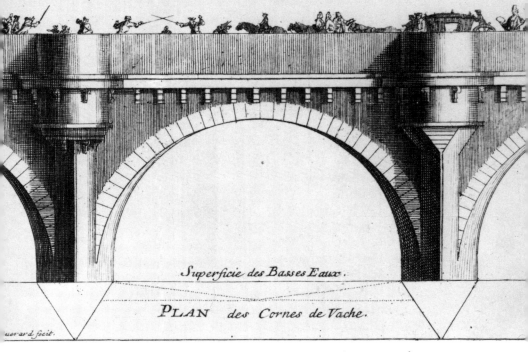

Superficie des Basses Eaux.

PLAN des Cornes de Vache.

uerard fecit.

A lively drawing *of the Pont Neuf in Paris, from Hubert Gautier's* '*Traité des Ponts*'.

work on the subject and remained so for nearly three-quarters of a century.

Growing consciousness of the need for trained engineers to further the work of the Corps des Ponts et Chaussées led, in 1747, to the formation of the École des Ponts et Chaussées and the appointment as its first director of Jean-Rodolphe Perronet. No greater engineer could have been chosen. He was born in 1708 at Suresnes, near Paris, and had been employed, first as an architect and then as an engineer. His work with the school led to his appointment in 1750 as inspector-general and finally in 1764 as first engineer of the Corps des Ponts et Chaussées, a post which he held with distinction until his death in 1794.

Perronet was one of the greatest bridge builders the world has known, and probably the greatest designer of stone arches. His

225

influence, not only in France but throughout the civilized world, can be traced through the works of his successors. His great bridge at Neuilly, crossing the Seine below Paris, had five arches each 120 feet in span with a 30-foot rise. The arch proper was approximately elliptical in form, being composed of a series of tangential circular arcs; the outer face, however, was segmental in form and the difference in curvature, increasing towards the abutments, was taken up by a bevelled inward slope. This feature was copied by Telford in his bridge across the Severn at Over, near Gloucester, which happily survives to demonstrate Perronet's idea as, unfortunately, the French authorities saw fit to replace Perronet's original in 1956.

His greatest surviving work, however, is carefully preserved—the Pont de la Concorde in Paris, acclaimed by many as his masterpiece; this is a matter of opinion, as others still consider that the Neuilly bridge represents Perronet's greatest achievement. The Pont de la Concorde was completed in 1791, after the outbreak of the Revolution and with that momentous event, the industrial initiative passed to Britain where the influence of the great French engineer and his contemporaries developed in an auspicious environment.

The earliest British pioneer was, however, a self-taught Welshman, William Edwards, whose bridge at Pontypridd is

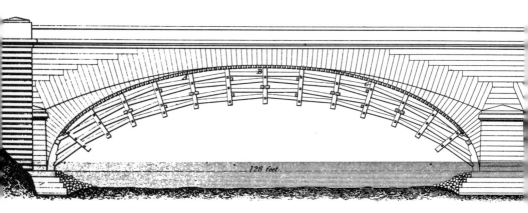

Perronet's stone bridge *across the Seine at Neuilly*.

Built to withstand *the violent flood waters of the River Taff, William Edwards' elegant bridge at Pontypridd was a triumph of perseverance and resourcefulness over the forces of nature.*

today a landmark representing, more than any other object, the transition of influence across the Channel.

The bridge at Pontypridd was described in 1803 by Benjamin Heath Malkin, whose comments are as effective as any made before or since. Malkin corresponded with David Edwards, William's son, and from these communications he describes the early life of William, who was the son of a farmer of the parish of Eglwysilan, Glamorganshire. Born in 1719, William Edwards was orphaned when two years old and became a rough mason employed in repairing the dry walls which were used on farms in those days. He studied the methods and tools used by visiting

masons and progressed to the building of sheds, cottages and ultimately a mill, on the construction of which he first became acquainted with the principles of the arch. Edwards' first bridge was carried away by a flood shortly after it was completed and he immediately undertook its replacement.

Benjamin Malkin describes it as follows:

'The second bridge was of one arch, for the purpose of admitting freely under it whatever incumbrances the flood might bring down. The span or chord of this arch was one hundred and forty feet, its altitude, thirty-five feet, the segment of a circle whose diameter was one hundred and seventy feet. The arch was finished, but the parapets were not yet erected, when such was the pressure of the unavoidably ponderous work over the haunches, that it sprung up in the middle, and the key-stones were forced out. This was a severe blow to a man, who had hitherto met with nothing but misfortune in an enterprise, which was to establish or ruin him in his profession. William Edwards, however, possessed a courage which did not easily forsake him, so that he was not greatly disconcerted. He engaged in it the third time; and by means of three cylindrical holes through the work over the haunches, so reduced the weight over them, that there was no longer any danger from it. These holes or cylinders rise above each other, ascending in the order of the arch, three at each end, or over each of the haunches. The diameter of the lowest is nine feet; of the second, six feet; and of the uppermost three feet. They give the bridge an air of uncommon elegance. The second bridge fell in 1751. The third, which has stood ever since, was completed in 1755.'

William Edwards' bridge remained in use until late in the nineteenth century, when another bridge of flatter profile but less elegance was built immediately alongside it. Edwards' bridge stands today as an ornament to Pontypridd and a memorial to the man of integrity who persevered in its construction.

During the latter half of the eighteenth century, the architectural tradition in bridge building flourished in Britain and produced a number of masonry bridges which followed closely the pattern set by the French. Among these were Westminster Bridge, built between 1738 and 1750 under the direction of Charles Labelye, a native of Switzerland. The middle arch was 76 feet span and, toward the abutments the spans progressively decreased to 25 feet. In 1760, a bridge to the design of Robert Mylne was begun over the Thames at Blackfriars. Mylne, who was an architect-engineer of Scottish extraction, was, from 1771, the surveyor to the New River Company, the first of three generations to serve the company continuously for 104 years. Blackfriars Bridge was built with nine elliptical arches, the centre one being 100 feet span, decreasing to 70 feet for the two adjoining the abutments. Both these bridges were built with the assistance of caissons, in which the lower courses of masonry were built before being sunk into the river on prepared piled foundations. Waterloo Bridge, built by John Rennie between the years 1811 and 1817, was, however, constructed with the use of coffer-dams, thus returning to a practice of the Romans. Rennie's Waterloo Bridge, with its nine equal semi-elliptical arches of 120 feet span,

Two 18th-century London bridges—*Westminster Bridge (top), and Blackfriars Bridge which was designed by Robert Mylne.*   229

remained to embellish London's river until it was replaced with the present reinforced concrete structure in 1945. London Bridge is still performing its function, carrying traffic far beyond any stretch of its creator's imagination. Designed by George Rennie under his father's direction, it was built after the death of the elder Rennie by John the younger (later Sir John Rennie). The five arches of London Bridge vary in span from the centre 152 feet to the 130 feet of the abutment arches.

Thomas Telford's stonemason origins naturally led to his masonry bridges being soundly practical, while at the same time, his early practice as an architect, combined with his study of the works of the French engineers, ensured that his works were always a pleasure to the eye. Telford's bridge over the Severn at Over near Gloucester was designed on the lines of Perronet's bridge at Neuilly. The arch stones of the outer faces of the bridge have the same chord as the inner arch, but the radius is much greater. The arch is therefore funnelled on both sides, the intention being to facilitate the passage of flood waters. This bridge represents Telford's nearest approach to a structural failure in his masonry bridges as, when the centering was struck, the arch sank 10 inches at the centre. This may be seen today in the line of the parapet and was due to a slip in the eastern abutment which led Telford to make the following comment in his autobiography:

Telford's bridge *across the Severn at Over, near Gloucester, was designed on the lines of Perronet's bridge at Neuilly. This bridge was Telford's nearest approach to a structural failure.*

Architecturally the finest *of London's bridges, Rennie's Waterloo Bridge, built between 1811 and 1817, was demolished in 1939.*

'Upon the whole, although the sinking of the large arch is small in comparison with what took place in M. Perronet's Neuilly Bridge, yet I much regret it, as I never have had occasion to state any thing of the sort in any other of the numerous bridges described in this volume; and I more especially take blame to myself for having suffered an ill-judged parsimony to prevail in the foundations of the wing-walls, leaving them unsupported by piles and platforms—because if so secured, I am convinced that the sinking of the arch would not have exceeded three inches.'

All these masonry bridges, technical and artistic triumphs though they were, merely extended the art of bridge building as already established. The real advance for which Britain was responsible was the application of iron as a principal structural material in the building of bridges and, from that, to other structures. Iron had already been considered by the French engineers as a structural material and, in 1755, iron arches for a bridge at Lyons were commenced, but proved too costly and the project was scrapped.

The first iron bridge was appropriately connected with the iron industry of Shropshire and was sponsored by John Wilkinson and Abraham Darby III. The ferry across the Severn between the villages of Madeley and Broseley near Coalbrookdale, had become inadequate for the heavy traffic resulting from the growth of the iron industry, and it was proposed to replace the ferry with a bridge. A Shrewsbury architect named Thomas Farnolls Pritchard prepared a scheme in which cast-iron ribs were used as permanent centering to a masonry bridge. A second scheme for the bridge included the basic idea of most of the later bridges in iron, being a series of cast-iron arch-shaped beams, pierced in the spandrels for lightness. This was superseded by a third design, the one adopted and still standing today, based on a series of five semicircular cast-iron ribs spanning 100 feet 6 inches with a rise of 50 feet. It is said that this design was largely influenced by Abraham Darby and his partner Reynolds, who, as expert foundrymen, saw to it that the design was feasible and who supervised its casting in open moulds. The casting of the main ribs of half the span was a major feat of foundrywork in those days. The bridge was built between the years 1776 and 1779 and the site is now named Ironbridge.

The original design *for this iron bridge over the River Wear at Sunderland was by Tom Paine, author of 'The Rights of Man'. It was erected in 1796.*

The world's first iron bridge, *still standing over 180 years later, was built across the Severn near Coalbrookdale by Abraham Darby, a local ironfounder. The site is now named Ironbridge.*

Other tentative efforts were being made about this time and at Merthyr Tydfil in Glamorganshire there is a cast-iron bridge which also has the appearance of being cast in open moulds to a design strongly influenced by timber construction. The iron of this bridge is reputed to be rust resistant.

Two cast-iron bridges of some note were completed in 1796, one by the revolutionary Tom Paine, author of the *Rights of Man*, who had designed it some years before, but fled the country before it was finally erected over the River Wear at Sunderland. This bridge was cast at Rotherham, exhibited in London, returned to Rotherham, and, finally, erected at Sunderland by Rowland Burdon as redesigned by Robert Wilson, the man responsible for its casting. In its final form it had a span of 236 feet and a low-water clearance height of 100 feet. The other bridge, of perhaps greater importance, was Thomas Telford's first effort in a material in which he later excelled, his iron bridge over the Severn at Buildwas. It was 130 feet span, little more than half the span of the Wearmouth bridge, but it was the forerunner of much greater works. Even while the Buildwas bridge was being constructed, Telford was working out ideas which were, at the beginning of the next century, to take shape in his monumental aqueducts at Chirk and Pontcysyllte.

Telford's greatest bridge, the Menai, was the final act of completion of his Holyhead road. Schemes for crossing the strait by arched bridges had been submitted by Telford himself and also by John Rennie, but these involved some interference with navigation and the Admiralty steadfastly insisted on that right being retained. In 1814, Telford designed a suspension bridge of 1,000 feet span to cross the Mersey at Runcorn, a project never carried out but which led to his undertaking a number of experiments in the strength of wrought iron bars, which served him well in his Menai masterpiece.

In deciding on his final design for the Menai Bridge, Telford was drawing on the experience gained in the use of iron for bridge suspension chains by Captain Samuel Brown, R.N., who was responsible for valuable inventions in chain cables, which led to their introduction into the Navy. In 1817, Brown patented the flat iron chain link which became extensively used in suspension bridges. Captain Brown designed the famous Brighton Chain Pier, which was built in 1823 and lasted until 1896, and the first large suspension bridge in Britain, of 361 feet span, over the Tweed, which broke up in a storm only six months later, probably due to the same kind of periodic oscillations which caused

Thomas Telford *from the oil painting by Samuel Lane. In the background is his great aqueduct at Pontcysyllte.*

One of the greatest works *ever achieved in iron*. Telford's bridge over the
Menai Straits carries the London to Holyhead road from the mainland to
Anglesey and was built high enough to allow the free passage of shipping

underneath. The design and construction of this great bridge were triumphs of Telford's skill, not only as a builder but also as an organizer.

the failure of the Tacoma Bridge over a century later. Telford's early experiments with suspension chains were carried out without knowledge of the parallel efforts of Captain Brown, but on hearing of the Captain's work, Telford collaborated with him and adopted the chain with flat bar links and pins, invented by Brown.

Following the many frustrations and delays brought about by numerous opposition interests, the Menai Bridge Commissioners, in 1819, obtained parliamentary powers to proceed with the bridge. By this date, the plans had reached an advanced stage and preliminary site work had begun. Telford had finally produced a design for a suspension bridge of 579 feet span between a pier on the Caernarvon shore and one on a rock near the Anglesey shore known as Pig Island. The headroom was the 100 feet required by the Admiralty. The approach viaducts consisted of three arches of 52 feet 6 inches span on the Caernarvon side and four of the same dimensions on the Anglesey shore. The bridge platform was 30 feet wide, divided into two carriageways of 12 feet with a central footpath of 6 feet and to carry this, Telford provided sixteen chains in four sets of four, each made up of five chain bars 10 feet long and $3\frac{1}{2}$ by 1 inch section; these had enlarged ends to link up with six chain plates at each end by means of two bolts. The deck was carried on suspension rods at 5-foot intervals; these were 1-inch square cross-section and hung from the chain joints which were staggered to suit the 5-foot interval. As with most suspension bridges built about that time, the deck platform was not sufficiently stiff to withstand the tendency to oscillate in a gale, and strengthening was needed soon after the bridge was put into service. Most of Samuel Brown's bridges failed for the very same reason, and the stiffness of the bridge platform is a major concern for the designer even of modern suspension bridges.

The Menai Bridge was not only a great triumph for Telford as an engineer, but it also demonstrated his genius for organization. In his preparatory experiments, performed with meticulous care, in timing the progress of construction, relating the delivery of materials to that programme and, most especially, in supervising

This revolutionary design by Telford *for London Bridge was to be carried out in cast iron. It was never built.*

the delicate operation of lifting the first chains, he displayed his ability to oversee works on the grand scale in a manner unexcelled to the present day. The bridge was opened at the end of January 1826, without ceremony, by the passing of the down Royal London and Holyhead Mail. The following July saw the opening of the less spectacular but still important bridge at Conway. The building of these bridges and of the great bridges of Robert Stephenson and Brunel were events, not only of technical interest, but of great drama, as Mr Rolt shows in his trilogy of biographies of these engineers.

The coaches had hardly begun to run on Telford's Holyhead road, when the railway age began. Soon, once more, the crossing of the Menai Strait offered a problem to Robert Stephenson who, in 1845, had been appointed engineer to the Chester and Holyhead Railway. For his crossing of the strait, Stephenson proposed to use a bridge of cast-iron, but the Admiralty were no more amenable than they were in Telford's day, so that any bridge requiring centering was out of the question. After considering various proposals, including the use of Telford's bridge, Stephenson decided on a crossing which aligned with the Britannia Rock in the centre of the strait, thus ensuring a good foundation for an island pier. His proposal was to erect a suspension bridge, but with a stiffening girder of depth and stiffness. This at first materialized in the design as an open-topped wrought-iron trough but it soon became evident that by closing the top, the girder would be greatly strengthened. To develop this idea, Stephenson enlisted the help of two men, each paramount in his own field; William Fairbairn, the mechanical engineer and expert on material testing, and Eaton Hodgkinson, F.R.S., the mathematician. Their work was based on a series of tests to destruction on large models with progressive improvements after each test and the end result, a self-supporting tubular girder bridge without supporting chains, was constructed in sections on platforms near the bridge site, floated into position, and jacked up to full height. Stephenson finally decided to build the approach spans of similar tubes, connecting them as a continuous girder with the river spans, and this was made effective by prestressing the girder by jacking down the end spans, a very advanced technique only recently revived. This continuity greatly increased the strength of the bridge, a fortunate circumstance today when trains of modern weights still pass on their regular schedules across the Britannia Bridge. The Conway river, also on the route, was crossed by a similar tubular bridge and this provided the builders with valuable experience for the erection of the greater structure across the Menai.

The first Conway tube was opened for traffic on 1 May 1849, and the second before the end of the year. On 19 June, the first of the great tubes of the Britannia was ready to be floated into position, but an accident led to postponement until 20 June when the great 472-foot tube was floated into position on the high tide, to be followed at intervals by the other three. By 5 March 1850, one line was completed and on that day, three locomotives towed a train of forty-five loaded coal wagons and with passenger coaches carrying a human load of 700 persons across the bridge. It was formally opened for single line traffic on 18 March 1850, and for both lines on 19 October in the same year.

While Stephenson's railways were extending in the north,

The work of two great engineers *can be seen in this view from Conwa Castle. On the left is Telford's elegant suspension bridge and, on the righ Robert Stephenson's tubular bridge carrying the Chester–Holyhead Railwa*

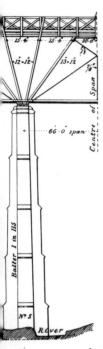

Ponsanooth
Viaduct (*de-*
tail) *is typical*
of many of
Brunel's stand-
ardized timber
viaducts in
Cornwall.

Brunel's broad gauge lines had been creeping westward and, in doing so, offered many bridging problems to the engineer. The deep valleys of South Devon and Cornwall were crossed by the most economical means then available, the beautiful, yet standardized, timber viaducts which only became uneconomic when the Baltic timbers of which they were originally built were no longer available. Have any of Brunel's critics tried in recent years to buy Memel timber in pieces of up to 60 feet length and 14 to 16 inches square? Yet such critics did not hesitate to express their views in support of concrete at the First Conference on Timber Engineering in 1961.

As with timber, Brunel was inspired in his use of wrought iron for bridges, inspiration based on hard work in experiment, testing, and further experiment, always reaching ahead. His wrought-iron girders, many still in use, put the material where it was most effective. His son's biography traces effectively the progress made through his bowstring trusses as at Windsor, to his Chepstow Bridge, a 'dummy run' for his final masterpiece, the Royal Albert Bridge at Saltash. His sketch-books, now in the library of Bristol University, indicate the development of his ideas leading up to Saltash and leave no doubt as to the thoroughly practical approach to all his problems. Beautifully drawn therein, Chepstow Bridge may be seen in the form which, until 1962, was a background to J. U. Rastrick's beautiful cast-iron road bridge, seen by the west-bound motorist as he approaches the traffic lights controlling the single line traffic crossing the Wye on Rastrick's 1813 bridge.

Brunel's great lenticular trusses now span the Tamar alongside the new suspension road bridge which has superseded the floating bridge of James Meadows Rendel.

The Chepstow Bridge, founded on cast-iron cylinders sunk under compressed air working, spans 300 feet by two trussed girders each formed of a circular tube of wrought iron as the upper flange supporting the rail platform as the lower flange by means of suspension chains. Saltash Bridge has two main spans of 465 feet each, with the centre pier founded in hard rock at a depth of 80 feet. The trusses are lenticular in shape, with the arch tubes rising to the same curvature as the opposite down-ward drop of the suspension chains. The tubes are oval in section, 16 feet 9 inches wide and 12 feet 3 inches high and each truss weighs over a thousand tons. As in the case of the Britannia Bridge tubes, Brunel's trusses were erected on platforms on the shore and floated out on pontoons, the first on 1 September 1857

The last journey of Rendel's 'floating bridge' across the Tamar at Saltash. It has now been superseded by the new suspension bridge which can be seen behind the trusses of Brunel's Royal Albert Bridge.

under the supervision of the engineer himself, and the second in July 1858, under the guidance of Brunel's right-hand man, R. P. Brereton. The bridge was opened in May 1859 by the Prince Consort, but Brunel was absent; he had spent the winter in Egypt on his doctor's orders and his only visit to the completed bridge was as an invalid. He died on 15 September of the same year.

With the completion of the Royal Albert Bridge, wrought iron as a structural material had reached its zenith. While the bridge was under construction, Bessemer had announced his new method for the manufacture of steel and, within the next five years, the new material was already being used for bridges. By the time the next great bridge was to be designed, steel had proved itself and, in 1880, Fowler and Baker had started their plans for the Forth Bridge.

The graceful suspension bridge *across the Avon at Clifton was designed by Brunel and completed after his death.*

A 'dummy run' for Saltash. *Brunel's tubular trussed railway bridge over the Wye at Chepstow is seen behind J. U. Rastrick's iron road bridge of 1813.*

The deep incisions into the east coast of Scotland formed by the Firths of Forth and Tay have always constituted important obstructions to communication by land along that coast. In reasonably calm weather conditions, ferries provided short cuts between places of importance on the north and south shores of each Firth, but in stormy weather, such crossings became hazards which could only be avoided by journeys of many miles by land. The idea of bridging the Forth was first suggested about the middle of the eighteenth century, at which time such a project was quite impracticable. In 1805, a proposal for a double tunnel under the sea bed across the Forth, each tunnel being about 15 feet wide and the same height, reached the stage of a prospectus inviting subscriptions for shares, but the project fell through. A more specific scheme was promoted in 1818 by an Edinburgh civil engineer, James Anderson, who prepared drawings for alternative designs for a suspension bridge having spans of 1,500 to 2,000 feet of 90 to 110 feet headroom, and a width of road plus footpath of 33 feet. The cost was estimated at £175,000 to £205,000 and the time of completion four years. Even at the prices ruling in those days the estimate was optimistic.

In 1860, the North British Railway selected a site between North and South Queensferry for the construction of a bridge to carry their line. Preliminary surveys and borings were made, a design prepared by Mr Thomas Bouch and an Act was passed in 1865. The scheme met with difficulties and was abandoned in favour of a proposal to establish a train ferry at the same place, which in turn proved impracticable. In 1873 Thomas Bouch designed a suspension bridge with two spans of 1,600 feet each and a company known as the Forth Bridge Company was formed by the four railway companies mainly interested in the traffic of the east coast; these were, the Great Northern, the North Eastern, the Midland, and the North British. An Act authorizing the construction was passed in the same year and a contract was let to Messrs W. Arrol & Co. of Glasgow. Work was started and extensive workshops laid out at Queensferry when, in December 1879, the Tay Bridge designed by Thomas Bouch failed. Public confidence in his ability was undermined by the results of an investigation into the disaster and the suspension bridge proposal was abandoned.

The railway companies interested invited their consulting engineers, Messrs Barlow, Harrison and Fowler, to consider proposals for the river crossing. This was done with great thoroughness and a number of alternatives, including tunnels, were investigated. The design recommended was for a crossing from a point about 6 miles west of South Queensferry to Garvie Island and North Queensferry by means of a continuous girder bridge of the cantilever and central girder type to a design which had been prepared by Messrs Fowler and Baker. This was approved by the directors and the Board of Trade, an Act being passed in July 1882. The working arrangement was for the Forth Bridge Company to erect and maintain the bridge, while the North British Railway Company undertook to maintain the permanent way and to manage all traffic.

The failure of the Tay Bridge had made engineers very conscious of the effect of wind pressures on structures, especially those in exposed places, and before any work was begun, a careful investigation was made into the wind pressures to which the bridge might be exposed. A value of 56 lb. per square foot, approved by the Board of Trade, was used in the calculations.

The steelwork for the bridge was fabricated in large workshops established on the South Queensferry shore. To these workshops were brought the raw steel in plates, rolled sections, and bars, and all the operations of bending, planing, drilling and so on, were done within view of the site of the bridge. Separate fabricating shops were used for the lattice tension members and the

A disaster in the annals of bridge building was the collapse of the Tay Bridge in 1879. A contemporary engraving shows the salvage operations on the following day.

A contemporary photograph *shows the building of the Fife cantilever of the Forth railway bridge. On the opposite page the new suspension road bridge is seen at an early stage of construction. The railway bridge appears in the background.*

tubular compression members, all the parts of which were clearly marked for assembly at the bridge itself. Erection required the use of much special plant, much of the riveting being done by hydraulic machines, with special travelling cages used for the riveting gangs. Many bridge building and civil engineering techniques, now used as a matter of course, were adopted for the first time on a large scale at the Forth Bridge. It may be well to

remember that, when the temporary works are dismantled and removed from any large work of civil engineering and the finished structure remains, those dismantled works often represent as much real engineering as, and sometimes more than, the permanent work.

The magnitude of the Forth Bridge contract naturally called for a large force of workpeople of all trades and grades. The employment of large numbers of men on contracts for railway similar works was not new, but the Forth Bridge brought these men into closer contact with each other and, at heights, into conditions of hazard both to themselves and their workmates. Accidents were frequent and, at times, led to strikes or unrest. Through this, the contractors made great efforts to improve the conditions under which the men lived and worked, although, according to modern ideas there was still room for improvement as, for instance, the contractor's contribution to the sick and accident club, which amounted to only 10 per cent or less of the total contributions by the men.

The Forth Bridge is almost invariably quoted when the subject of maintenance of structures is discussed. The painting of the bridge is, of course, a continuous operation and it is interesting to note the careful provision made for protection by the first painting specification. Before erection, all the steel components of the bridge were scraped, wire brushed and coated with boiled linseed oil, applied hot. The next coat, applied either before or after erection, was of red lead, and a second coat of red lead was put on after erection. Over this, a priming coat of oxide of iron paint, dark chocolate-brown in colour preceded the finishing coat of the same kind of paint in a bright Indian or Persian red. The inside of the tubes received one coat of red lead and two coats of white lead paint. In all, the calculated area of steelwork to be painted adds up to a total of 145 acres.

The first test trains, of 900 tons each, crossed the bridge on 21 January 1890, and the engineers principally concerned with the work received the honour due to them, John Fowler's including a baronetcy, Benjamin Baker, a knighthood, and William Arrol, the principal contractor, also a knighthood.

Trenton Viaduct *on the Delaware River affords an example of the many Pratt trusses built in iron for railways in the U.S.A.*

While the effects of the industrial revolution influenced bridge design in Europe, the comparative expense of iron in America led to a far greater use of the abundant timber resources of that country and with it, great ingenuity in the design of truss bridges. Many variations of truss forms were devised, such as the lattice truss patented by Ithiel Town, a New Haven architect in 1820. This truss, formed of planks nailed at the intersections, may fairly be considered the prototype of the great number of iron and steel lattice girder bridges used later in the century for railways in all parts of the world. Although they were not objects of beauty, they were economical to build and easily fabricated at remote sites, and thus furthered the rapid expansion of communications in remote parts of the world.

Timber trusses were usually designed to use the principal material in compression, with iron rods as the tension members. Such designs were uneconomical when iron alone became the material of the whole bridge and, in 1844, Thomas Willis Pratt, with his father Caleb, patented a more economical truss which, like its British equivalent designed by James Warren, has served its purpose for medium spans up to the present day.

The longest single span *in the world (4,200 ft) is seen in this photograph of the Golden Gate Bridge in San Francisco.*

Concurrently with the development of truss design, American engineers have consistently advanced the design of suspension bridges. An outstanding figure in this respect was James Finley who, 25 years before the opening of Telford's Menai Bridge, built his first chain suspension bridge. Finley's eight suspension bridges, built between 1801 and 1810, were all successful. In days when the laws of aerodynamics were completely unknown, this man, by the exercise of his engineering art, devised the first stiffening girders in the form of substantially braced railings, thus preventing the build up of those wind-induced oscillations which have caused the failure of so many suspension bridges before and after his time. Great attention is given by modern designers to the

design of stiffening girders for suspension bridges and James
Finley may be said to be the first engineer to appreciate the
importance of this.

American engineers have also excelled in the use of wire cables
for bridge suspensions. Although the principle was originated in
France by Louis-Joseph Vicat, whose name is commemorated in
cement testing, the fullest development of wire suspension
cables must be credited to John Augustus Roebling who, in 1844,
developed a method of spinning these cables of continuous wire

Brooklyn Bridge, *over the East River in New York, was the first
of a great series of wire cable suspension bridges and the greatest achievement of J. A. Roebling.*

which has substantially remained the standard practice up to the present day. First used at Pittsburgh, he later crossed the Niagara gorge in 1855 with a bridge that survived until dismantled in 1896 (see p. 125). Roebling's greatest achievement, however, was his Brooklyn Bridge over the East River in New York. Of 1,535 feet span and 133 feet clear height over the river, it was completed after his death by his son W. A. Roebling. The Brooklyn Bridge was the first of a great series of wire cable suspension bridges, the construction of which is continuing up to the present day.

Since the Forth Bridge, steel has maintained its position as a material for bridge construction, but, even as the first trains crossed the estuary, a rival material was coming into use, which was to become equally, if not more important—reinforced concrete. The early use of reinforced concrete for bridges showed a tendency on the part of their designers to imitate steel construc-

Two concrete bridges of the 20th century—Sando Bridge (left) over the Angermann River in Sweden and the Narrows Bridge, Perth, Western Australia (below).

tion, but designers such as Hennebique realized its potential and produced bridges of symmetry, using to the full the qualities of this new combination of concrete and steel. The full value of this combination was not exploited until, in the second decade of the present century, Freyssinet made full use of the compressive strength of good concrete by the practice of prestressing.

Although the principle had been known and discussed since the mid-nineteenth century, the practical application had to wait until reliable high tensile steel wire could be produced in quantity, and designers had devised means of anchorage for the tensioned wires which form an essential part of the construction. Today, prestressed concrete is an accepted form of construction enabling engineers to design bridges which are both economical and beautiful.

The Medway Bridge has a centre span of 500 ft and a width of 113 ft 6 in, making it the largest prestressed concrete span built up to 1963.

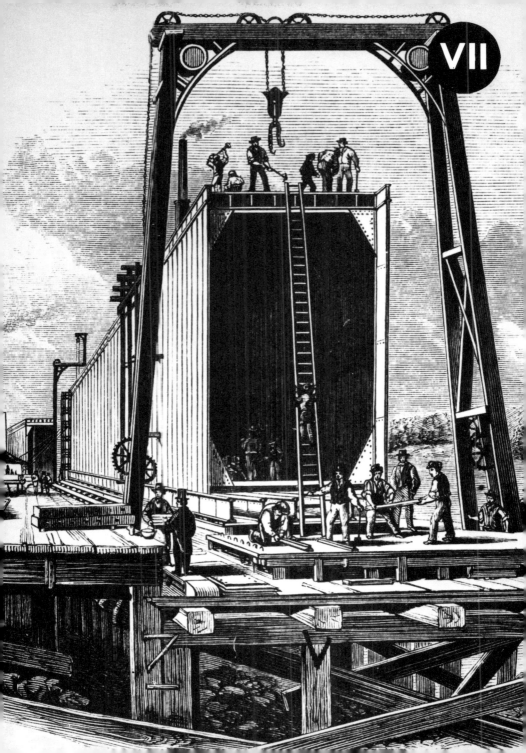

VII

Earliest materials — brick, timber and stone

Roman cement — the Middle Ages — theory of structural

stability — discovery and working of iron — coal in

ironmaking — testing of materials — Hooke, Coulomb

and Reamur — Abraham Darby at Coalbrookdale

Henry Cort — Penydarren and Trevithick

Fairbairn and Hodgkinson develop systematic testing

Bessemer's converter process — bridges of steel

modern cement-making

◄One of the main tubes *of the Britannia Bridge, built by Robert Stephenson across the Menai Straits, being erected on the staging. The tubes were floated into position on pontoons.*

Mastery of Materials

ACCORDING TO THE evidence available for study by the archaeo-logist, the earliest materials used by man for works of construction were earth, stone and timber. Earthworks of great magnitude were carried out with simple tools of bone, wood, and flint; no doubt vast quantities of soil were transported in simple con-tainers such as baskets holding one or two cubic feet. Stone would have first been used as found in its natural state and timber would have been shaped by fire or, later, with flint axes and scrapers. The plastic properties of mud were recognized at a very early stage in man's development and, especially where the mud was composed of suitable mixtures, its use as a cement or plaster enabled stones to be used more advantageously. The practice of moulding mud and clay into bricks had developed by 4500 B.C. so that large structures were by then possible, and not only houses but buildings for communal or civic use came into being. The bricks were made of mud or river clay, worked up with fibrous material such as straw, reeds, or grass for improving the mechanical properties and to prevent crumbling, and were then dried in the sun. The sun-drenched, fertile valleys of the Nile, Tigris and Indus all lent themselves admirably to the requirements of brick-using peoples, with the mud freely available and abund-ant sunshine for the baking. It was, however, in the valley of the river Jordan that the earliest brick buildings yet found by arch-aeologists were built. In its deep valley, 1,200 feet below sea-level, the river winds through swamps of infertile mudland; but, at one place only, the sweet water springs up and forms a rich oasis and here 9,000 years ago, men of the Neolithic period built the first town of Jericho. These people probably lived in tents, but by the fifth millennium B.C., the inhabitants of the town had come to build houses of bricks, not bricks as we know them, but nevertheless, hand moulded and baked flat pieces of clay made into huts of a beehive shape. Surrounding the whole city was a massive stone wall, the earliest town wall so far discovered, and proof of an organized community defending itself against its enemies.

The baking of bricks was not always considered necessary; buildings have been found in which the bricks were laid moist as they came from the mould. Usually, however, bricks were dried and bedded in sand or mortar, while, at a very early date, bitumen was used as a bonding agent. There is no positive date known when the practice of burning bricks was introduced, but in Egypt not only burnt bricks but glazed tiles, used as a facing material, provide evidence of kiln burning.

Timber was a very scarce material in the scorched lands of the Middle East, so that its use for temporary works was avoided. Brick arches, which were used in Egypt and Iraq from very early times, were constructed by methods designed to avoid the use of centering. The successive courses were corbelled out as canti-levers until they met, or the bricks were laid in arch rings of independent slices, relying on mortar to bond adjoining slices to each other. The vaulting of the great hall at Ctesiphon, built in A.D. 550, was constructed by this method to a clear span of 86 feet, with a rise of 105 feet.

The Greeks used brickwork only in conjunction with timber, in their earlier days as a filling to timber frames, and later as mass work with timber bonding. Up to the time of Augustus (27 B.C.) the Romans used unburnt brick and, in Imperial times, their bricks were burnt, but in no case did they use brick as a principal material of construction. The Persians used brickwork for important buildings which developed into a style, known as Byzantine, which spread to western Europe along the trade route across Asia Minor, Greece, Venice, Lombardy, the Gulf of Lions and the Garonne Basin to the Atlantic coast. Persia also influenced the building practice of the Mohammedan countries, whose distinctive style of building spread through Syria, Egypt and North Africa to Spain.

Byzantine art in building culminated in the great cathedral church dedicated to Hagia Sophia, the Sacred Wisdom, which

Byzantine art *in
building culminated
in the great cathedral
church of Hagia
Sophia in
Constantinople,
built about* A.D. 530.

*The brickwork
detail (above) is of
an arch in the
palace of Ctesiphon.*

was built in Constantinople about A.D. 530 approximately contemporary with the Ctesiphon palace. It had a dome 107 feet in diameter formed of brick ribs each 2 feet 4½ inches wide at the springing. The ring on which these ribs rested was carried by pendentives and by four great arches to four corner piers, the outward thrust being taken at the sides by bastions, and at the ends by apses. The principles of its design are a mystery; at the time there was little or no science of mechanics on which to base calculations and, from an early stage, trouble developed. The piers and counterforts began to spread. It became necessary to add weight by filling in some of the ornamental niches, to enlarge the side bastions and the haunches of the cupola; the staircases were reduced in width, but all these modifications were not sufficient especially as the church was subjected to earthquake shocks which aggravated the settlement and spread. The dome was therefore completely rebuilt with a greater rise and has subsequently been partly rebuilt several times. The form of the building is a strictly functional brick carcase, with its glory depending on decoration in marble, mosaic, enamel and gold, and as such it represents the peak of achievement in construction and decoration of the great Roman and Oriental builders.

Brick construction is one of the arts which successfully survived the fall of the Roman Empire and, from North Italy, it spread to north-west Germany and the Low Countries, where domestic architecture of great beauty developed through the use of brickwork. By the fourteenth century it had crossed the North Sea to East Anglia, one of the arts introduced to Britain by settlers, often political or religious refugees, from the continent of Europe. The many clay areas in Britain encouraged the use of brick, so that in those areas it largely superseded the use of timber for exterior work in buildings and, except where stone was a local and cheaper product, that material also, except for building work of a prestige nature. Although the Netherlands and West Germany have the greatest number of examples of beautiful brick building dating from early times, Britain has many, especially from the Tudor period onward. Hampton Court is probably the best example, incorporating, as it does, the sixteenth-century work

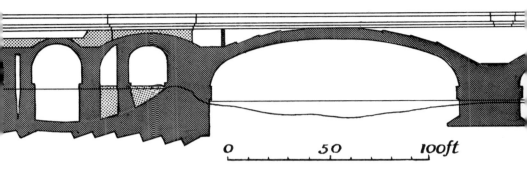

The long flat arches of Brunel's Maidenhead Bridge were a masterpiece of brickwork.

in Wolsey's palace, with bricks only $1\frac{1}{2}$ inches thick, and the additions in the eighteenth century by Sir Christopher Wren.

The economical use of local materials is usually an essential requirement in the design of building and civil engineering works, and it was natural that the civil engineers of the canal and railway periods should turn to brickwork as a main material for their works in the extensive clay areas of western Europe. In such areas, not only the raw materials, but skilled bricklayers were available in great numbers, and this led to the development of harder, stronger bricks for use in engineering. Bridges, viaducts, retaining walls, tunnel linings and similar works could be erected in brick to adequate standards of safety and in close approximation to estimated costs. Brunel, in particular, saw the potentiality of brickwork and used it extensively in the brick areas of the Great Western Railway, achieving that masterpiece of brickwork, Maidenhead Bridge, the longest, flattest brick arches ever achieved.

Stone is such a natural building material that its use dates back to the remotest periods of man's activity as a builder. The forces of nature provide, in many parts of the world, stone in sizes and shapes which lend themselves to rough construction with little

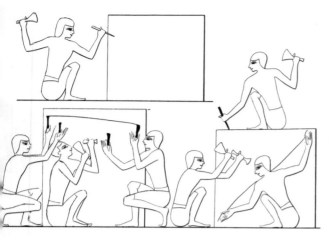

Copper chisels *and round-headed mallets are used by ancient Egyptian stone masons to square up a block, possibly for a sarcophagus.*

or no further preparation. Laid on each other, or with mud, clay, or earth to fill the voids, such stones have been, and are still being used for walls and similar purposes, probably for a greater period than any other building technique and are more widely spread geographically; so that excavations in the Middle East, India, China, and the Western Hemisphere demonstrate that, as with other inventions, the development of stone as a building material was inevitable. It was therefore natural that as man came to use tools, he proceeded to alter the shapes of his stones as a means of improving their mutual stability and to enhance the appearance of his structure. No doubt, such cutting was first done to the softer stones, but, by the time of the Egyptian civilizations, it had become possible to work the hard granite, syenite and basalt available in the southern part of the country.

As with other ancient communities, time in Egypt was no deterrent and labour was plentiful. The Nile provided the transport and also the labour as, during each flood season, vast numbers of valley dwellers were driven from their homes and it is probable that through this the practice began of providing relief for the unemployed by public works. Various methods were adopted for extracting the rock from the quarries; low grade surface rock was heated by fires and quenched with cold water, probably a dangerous experience for the quarrymen, but labour was expendable and, in any case, many of the projects had a religious or patriotic significance, fostered by the priesthood, which encouraged zeal in the workers. In any case, all reserve foods were held by the rulers, so that it was a matter of work or starve. Other methods, equally or even more laborious, included forcing the stones out by driving dry wood wedges into fissures, the wedges, on being wetted, then expanded and burst out the stones. If the rock was not fissured, heavy balls of dolerite were used to pound out a trench on both sides of the stone, which was then undercut with emery abrasives embedded in copper tools. The pounding process was used to bring the stone to its approximate shape when it was finished with the same copper-emery combination, or, at a later date, with work-hardened copper chisels and the round-headed mallet used by masons for so many centuries since.

Using such methods, the Egyptians raised monuments which remain to this day, as they were seen by the ancient Greeks to be among the wonders of the world. After the Egyptians, the use of stone declined and, except for some decorative purposes, was replaced by brickwork until about 2000 B.C. when stone again became the material for buildings of importance. By about 700 B.C., however, the Greek civilization had developed architecture

Repairs *to the Parthenon steps are carried out by modern Greek masons.*

to a high form of art, using stone as the principal medium of construction. The early Greek buildings in stone were designed in a tradition based on the use of timber. Column and lintol forms were predominant, and the timber traditions even extended to the shape of columns, which, during the Mycenaean period were of a tapered form, having the largest diameter at the top. This practice derived from the nature of timber posts which, if buried in the ground, had a longer life if erected root end up, the larger diameter at the root end also offering advantages in fixing the roof timbers. Such practices based on timber technology took several centuries to work out of the system. The persistence of this feature in columns of stone may have led to the acceptance of tapering columns with the larger diameter at the top as being aesthetically desirable, but the theory that they were deliberately designed to counteract the effects of perspective when looking upward appears to be of modern origin.

The Greeks had the benefit of iron as a material for tools. They also had uniformly good limestone and marble to work, so that, with the minimum of labour, they were able to equal, and later surpass, the best work of the Egyptians. Their combined use of the services of artists and mathematicians enabled them to produce forms of great beauty with functional stability, many of which have remained prototypes for architects up to the present day.

The art of the Greeks persisted throughout their works, but the Romans were of a different mind. Strict utilitarianism is the keynote of all Roman work and the forms, based on stability, are rugged and simple, ornament being only incidental, and added almost as an afterthought. The Roman Empire, based on military conquest, represents the first, and probably the greatest, example of a military organization based on technology, and that technology was civil engineering. The geology of the country around Rome lent itself to considerable developments in the practice of building in stone. The first city was built largely of mud bricks, but in turn, the Romans developed the use of tufa, a volcanic ash conglomerate, easily workable; perperino, a harder conglomerate; travertine, a hard crystalline limestone; and later, marble for important works. Stucco was developed for external finishes to

The impressive size of Roman brick construction can be seen in this series of arches on the Palatine Hill.

works of intermediate importance. The greatest asset of the Romans, however, in their building, was their cement. The secret of this was lost after their times until recent years, but its properties depended on the inclusion of a proportion of pozzolana, an earth of volcanic origin, which, when mixed with lime, produced a cement with hydraulic properties, that is, it would set hard in water, and concrete made with this cement in Roman times remains sound to the present day.

The masonry of the Romans varied according to the stone available, the skill of the masons—although this was usually of a high order—and the purpose of the structure. Hard stones were used for structures needing strength, such as columns, piers, and

267

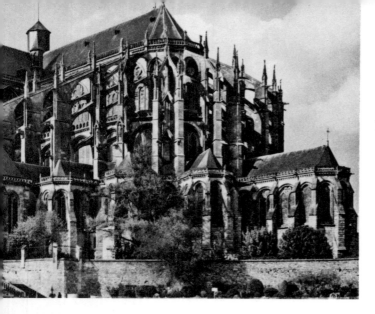

Function and form *combine in the flying buttresses of the Gothic cathedral at Le Mans.*

arches; softer stones were used for interior walls. Squared stones were often used for face work to a core of concrete consisting of random stones in mortar, the Roman work in this respect being much superior to that of the mediaeval builders who used little or no mortar in the rubble core behind their facings. Ironwork was used where the builders thought it necessary to strengthen the natural bond, lead being used to secure the iron to the stone. The Roman arch was invariably circular and was built on centering, for which corbels were provided in the piers below the springing. The arch and spandril facings provided a mould into which the concrete was placed in layers consisting of a mortar matrix filled with stones rammed in. Fortress walls and other works of utility were built with random stones in mortar, often laid in herring-bone pattern and usually with string courses of brick at intervals to maintain the horizontal alignment.

The vast works in masonry left by the Romans became quarries for many buildings of subsequent centuries; these, in their turn, often became the source of material for still later generations to build with, stone being such a durable material, and, once wrought to rectangular shape, so convenient to re-use. Walls built with Roman cement must, however, have been much harder to demolish than those built in later centuries with ordinary lime mortar.

The early builders of the Middle Ages followed Roman practice as far as their skills would allow, and thus architecture in western Europe was solid and heavy. The Arabic people, however, tended towards higher buildings with pointed arches, which their more advanced technology had achieved. These ideas gradually filtered through Europe from the time of the Crusades, the influence being seen first in France, then in Britain and finally in the Rhineland, the whole transformation taking about a century. As the fraternity of masons gained experience, so their ecclesiastical employers became more ambitious. The collaboration of the two inspired the erection of buildings in stone which stretched to the full those properties of stone which the masons had become so expert in cutting.

The Gothic period, as this great era of architecture became known, started to wane in Italy during the fifteenth century, in France during the sixteenth, and in England in the seventeenth. This was due to a number of causes. There was new interest in classical forms, arising from the re-discovery of the works of Vitruvius, whose book on architecture opened up a new vista when it was printed in 1485. The Gothic forms had been developed to the limit of safety; central authority had become powerful in most States, commanding vast resources, and demanding official, commercial and private buildings of a size commensurate with the importance of the new pattern of administration, no longer ecclesiastical.

The new interest in scientific investigation led, not only to the theory of structural stability, but to the practical testing of materials. One of the first to undertake such testing was Robert Hooke, who, in the mid-seventeenth century, described his experiments in his paper 'Of Spring'. Mariotte also, at about the same period, was advancing theories based on tests with wooden and glass rods. The eighteenth century, however, brought greater advances in material testing, partly through the work of Coulomb, whose work on the strength of materials was the fruit of a series of experiments on test pieces. Réaumur also made tensile tests of wire to evaluate the results of heat treatment and made elementary tests of hardness. The first real testing machines on modern lines were constructed by Peter van Musschenbroek (1692–1761) professor of physics at Utrecht and later at Leyden. This machine, although it was only powerful enough to test very small specimens, nevertheless, was adaptable to tensile, compressive, and bending tests, and his results, published in his book *Essay on Physics* printed in Leyden in 1751, were used extensively by the engineers of those times. Since Musschenbroek made his tests on very small specimens, the test pieces were too small to be reliable and Buffon, the director of the Paris botanical gardens, was critical of Musschenbroek's work. He produced results of his own, based on tests of large sections loaded as beams and these confirmed Galileo's statement that the strength of a beam is proportional to the width of its cross-section and to the square of its depth. They also showed the proportional variation of the strength of beams in relation to the density of the timber.

Buffon's work on timber was followed by similar tests of stone by the French engineer Gauthey who, using a lever machine, tested specimens 5 centimetres square. This machine was improved by Rondelet who introduced the knife-edge principle into the design of testing machines. By the end of the eighteenth century, French engineers worked from tables of the strength of iron, timber, and stone, produced by experimenters such as Gauthey.

Until the end of the seventeenth century, iron had never been used as a principal structural material, although it had, for many centuries, been an important means of fastening structural components in other materials, especially timber. The discovery and working of iron was a major step in the history of mankind, but great resources of skill and knowledge were needed for its exploitation on a large scale; it remained a scarce material to be used sparingly and for mainly warlike purposes. Iron was worked by the Egyptians about 1500 B.C. or even earlier, for making into weapons, nails, and other small articles and, from the Egyptians, the Greeks learned the art of producing not only iron, but possibly steel. By the time of Alexander the Great (355 B.C.), Daimachus, a contemporary writer says:

The Roman blacksmith's tools *were much the same as those used
by village blacksmiths today. The metal is heated with the help of
bellows in the furnace and then held with the tongs to be hammered.
The tongs, hammer and chisel are seen on the right.*

'of steels there is the Chalybdic, the Synopic, the Lydian,
and the Lacedaemonian. The Chalybdic is best for carpenters'
tools, the Lacedaemonian for files, drills, gravers, and stone
chisels; the Lydian also is suited for files, and for knives, razors,
and rasps.'

This classification seems to indicate differences in local practice
in the art of tempering, as carpenters' tools are taken to a lower
degree of temper than the other implements referred to. Steel
also may have been used by the Greeks for armour as Pope's
translation of Homer, in referring to Agamemnon's armour, sets
forth that: 'Ten rows of azure steel the work enfold,' but this
interpretation may be due to Pope, not Homer. The manufacture
of iron and steel was well advanced in India at about the same time
as Aristotle in 384 B.C. describes the making of the Indian 'wootz'
steel in the following way: 'It was produced by heating on a charcoal hearth about 1 lb. weight of malleable iron, cut into small
pieces, with about 10 per cent of dried wood, in clay crucibles,
the covers of which are luted on with clay.'

271

The Romans manufactured iron and steel in considerable quantities, as the extensive use of the metal by their armies demonstrates. It was also used extensively for domestic hardware such as nails, bolts, hinges, locks, knives, scissors and hunting weapons. Their introduction of its manufacture into Britain led to the establishment of iron industries in districts which remained centres of iron making until the prohibition of iron making by charcoal in the seventeenth century. The chief of these centres were in Sussex (the Forest of Anderida), the Forest of Dean, and in Yorkshire. In these districts, cinder beds and other evidences of metallurgical industry extend for miles. The art of making iron does not appear to have entirely died out after the Roman Empire fell and, apart from the countries of the East, which were unaffected by that event, centres in Europe continued to make the raw material of weapons. At the time of the Norman invasion, Gloucester was a centre of the iron trade and the Domesday Book indicates that tribute of that metal, in the form of bars, exempted the city from any other tax. An Act of Henry III prohibited the export of iron and regulated its price. The home industry was given an infusion of new experience at that time by the settlement in England of German steel-workers.

By the sixteenth century, iron smelting had made such inroads on the forests which supplied the charcoal that the supply of timber for shipbuilding was prejudiced. In 1558 and 1581, Acts of Parliament restrained the working of iron in certain areas. In spite of this, the industry flourished in other parts of the country, especially in the Midlands, where growing timber was too far from the coast to be of use for shipbuilding. The prosperity of the iron and steel trade at this time attracted skilled workers from many parts of the Continent and the quality of the steel produced improved accordingly. Methods of improving the air blast were introduced and iron, from being only barely workable in a plastic state except in small quantities, could be heated to a fluid and cast into moulds. The art of gun founding, which originated in France and the Low Countries, was brought to England and its importance outweighed considerations of timber conservation so that iron working became re-established in Sussex. Not only were cannon cast, however, but the new technique enabled iron to be used in the cast form for many domestic purposes, such as cooking pots, fire backs and grates. The industry prospered so greatly in Sussex that, by the end of the sixteenth century, cannon and other iron goods were being exported in some quantity, and it was estimated that, during the reign of James I, of the total of 800 iron mills in England and Wales, 400 were in Surrey, Kent and Sussex. Although this figure may have been exaggerated, the Sussex iron industry was, by that time, causing great concern to those interested in the building of ships and the Act of 1581 was vigorously enforced. Many of the prosperous Sussex ironmasters moved to South Wales and, in Glamorganshire, set up their forges in the districts of Aberdare and Merthyr Tydfil, where iron ore and timber both existed in abundance. Before the end of the eighteenth century, no ironworks remained in Sussex and the industry was established in those parts of the country where the next technical step became inevitable, that of using coal for smelting.

Early iron smelting made such inroads on the forests which supplied the charcoal that the supply of timber for shipbuilding was affected.

The early history of the use of coal for making iron is confused. The wording of early patent specifications was intentionally vague, and inventors made a great mystery of their processes. In the early seventeenth century, Simon Sturtevant, a German, took out a patent 'to neale, melt, and worke all kind of metal oares, irons, and steeles with sea-coale, pit-coale, earth-coale, and brush fewell'. This invention may have been intended to cover the coking of coal, but as none of the process was described clearly, and the production of iron did not follow the granting of the patent, the rights were cancelled after a year. Other patents followed on the same lines, and no doubt some of the patentees knew what they hoped to achieve—the production of a fuel from coal which would satisfactorily replace charcoal, but none succeeded in producing iron in commercial quantity. In 1620, a patent was granted to Lord Dudley on behalf of his son Dud Dudley, 'for melting iron ore, making bar-iron, etc. with coal, in furnaces, with bellows'. Dud had a chequered career during the time of the Civil War on account of his royalist activities, and there is no doubt that he built up at one time a flourishing business in the production of iron, but the secret of making it with coal, if he ever possessed it, died with him in 1684.

The first successful foundryman to use coal in the form of coke was Abraham Darby, the first of a succession of ironmasters bearing the same name, who opened up a business at Bristol in 1700, to work at his trade of making malt-kilns, which he soon expanded to include brass and ironfounding. The purpose of the venture into iron was the great demand for iron pots which, at that time, were imported from Holland. Darby tried to mould his pots in clay, but was not successful, so he went to Holland in 1706 and there found the simple secret of successful ironfounding —moulds of fine, dry sand. He managed to patent this secret in 1708 and prepared to embark on the large scale manufacture of iron ware. His partners, however, refused to invest more capital in the business, so Darby severed his connection with them and in 1709 went to Coalbrookdale, in Shropshire. This district had been a centre of the iron industry from Tudor times, but shortage of fuel had closed down the furnaces. Darby's coming started a revival which was to make the name Coalbrookdale famous throughout the world. At first Darby used charcoal in the traditional manner, but the shortage of this fuel led him to use coke, which he prepared from selected coal and, with the help of more powerful furnace blast than used by earlier workers, was able to rely entirely on coke fuel. Abraham Darby the first died in 1717, leaving a flourishing ironfoundry, which his sons were too young to work and which was sold at a great loss.

In due course, the next generation of Darbys grew up in Coalbrookdale, and took up their father's trade of ironfounding under the management of Abraham the second.

In 1763, Abraham II was succeeded in the management by Richard Reynolds, who was born in Bristol in 1735 and, like the Darbys, was a Quaker. Reynolds married the daughter of Abraham Darby and took over the works management on his father-in-law's death. He became the manager of a large and flourishing business, with foundries, not only in Coalbrookdale, but also in London, Bristol and Liverpool, and agencies at Truro and Newcastle for the sale of mining machinery, including Newcomen's pumping engines. Reynolds greatly improved the ironfounding processes, but, of greater importance at that time, he introduced the use of coke in the manufacture of wrought-iron bars. This innovation was suggested to him by the brothers Cranage; Thomas, of Bridgnorth; and George, of Coalbrookdale. These two experienced iron-workers, in 1766, suggested to Reynolds that coke might be used for the process of refining pig iron into malleable iron by separating the fuel from the iron, using what became known as a reverberatory furnace with the metal in a separate furnace chamber. Reynolds experimented with a small furnace, which was highly successful, enabling him to convert the pig iron to good malleable iron capable of being drawn out under the hammer into all shapes and sizes. The process was patented in the names of the Cranages in 1766 and from that time wrought malleable iron became available in greatly increased quantity at a much lower price—a stimulus to industry at a time when great things were being done.

One of the results of the new process was the conversion by

Coalbrookdale in Shropshire was a centre of the iron industry from Tudor times. This iron works may have been that started by Abraham Darby in 1709.

Reynolds of all his tram roads to iron rails. This was done in 1767 and from this conversion ultimately developed the many thousands of miles of railway track which were to spread all over the land surface of the world.

The iron made by the Cranages' process at Coalbrookdale tended to have excessive proportions of carbon and sulphur which made it 'red short', that is, it crumbled when being worked hot under the hammer. The same difficulty was met by Peter Onions, of Merthyr Tydfil, who, in 1783, obtained a patent for a puddling furnace, or one in which the iron maker could stir the metal while it is kept in a plastic state by the heat from a fire. The puddling

or stirring process was continued until the carbon was reduced to a minimum, when the ball or lump of iron was raised to a white heat, removed from the furnace, and forged under the hammer.

The man whose work did most to put eighteenth-century iron making on a sound footing was Henry Cort, who was born near Lancaster in 1740 or 1741. In his early years he became a Navy Agent or banker-paymaster for the distribution of pay, allowances, prize money, etc., for the Admiralty. His office was in London, but his duties brought him into contact with the naval ports. In 1775, as a result of the financial embarrassment of the owner, a Mr Morgan, Henry Cort, as principal creditor, took over the management of an iron works at Fontley, near Titchfield, in Hampshire. This forge had existed from the early seventeenth century, or earlier, and had a tilt hammer worked by water power from the River Meon which, when Cort took it over, produced about 200 tons of iron per annum. Fontley was admirably situated for the manufacture of iron to meet the requirements of Portsmouth Dockyard. In the midst of wooded country, it communicated with Portsmouth and Gosport by Fareham Creek, which was navigable at high tide and, as the business developed, Cort opened up a works and store at Gosport with wharf access to the harbour. He entered into large contracts with the Navy for the supply of iron hoops, a necessary part of the construction of wooden masts and spars, taking as part payment the old iron hoops which had previously been sold as scrap.

To make these contracts pay, Cort experimented with improved methods of producing iron and of forging it into flat bars. This involved him in considerable expenditure and to help him in this, he took into partnership Samuel Jellicoe, the son of Adam Jellicoe, who had been associated with him in naval pay transactions and to whom Cort became greatly indebted.

Henry Cort's great contribution to iron technology was to take the many, partly successful, methods of his predecessors, and to develop them, with his own improvements, into an efficient and practical iron making procedure which could be expanded in scale to produce great quantities of wrought iron at a low price. His first patent, of 17 January 1783, described how he took the old mast hoops or any other old iron bars and after heating in a reverberatory furnace, fed with ordinary pit coal, he folded and bundled them into faggots; several of these were then brought to a welding heat and forged under a tilt hammer of 8 or 9 cwt. Large iron pieces were handled by welding 'porters' or handling bars on to them. After the faggots were welded under the tilt hammer, Cort passed the iron through a rolling mill to squeeze out the cinder, re-faggoting and repeating if necessary to improve the quality, and finally passing the iron through grooved rollers to produce the sections required.

An efficient *but cheap method of producing large quantities of iron was Henry Cort's outstanding contribution to iron technology.*

The demand by the Navy Board for iron became so great that Cort was persuaded by them to try to convert old cast-iron ship's ballast or kentledge into wrought iron. His next patent, dated 13 February 1784, describes one of the greatest advances in iron making, the process of dry puddling. Cort's description, unlike most patent specifications of those days, is clearly worded and leaves no doubt as to the procedure adopted. He heated the metal in a reverberatory, or air, furnace fired with raw pit coal. The iron was either melted on the hearth, or was brought from the blast furnace in ladles. Apertures in the furnace doors enabled the iron maker to stir the contents from time to time. The patent specification continues:

The iron is forged
*under a tilt hammer
in this 18th-century
painting by Joseph
Wright of Derby.*

'After the metal has been for some time in a dissolved state, an ebullition, effervescence, or such-like intestine motion takes place during the continuance of which a bluish flame or vapour is emitted and during the remainder of the process the operation is continued (as occasion may require) of raking, separating, stirring and spreading the whole about in the furnace till it loses its fusibility and is flourished or brought into nature. . . . As soon as the iron is sufficiently in nature it is to be collected together in lumps called loops, of sizes suited to the intended uses and so drawn out of the door or doors of the furnace. . . . The method and process invented . . . by me, is to continue the loops in the same furnace or to put them into another furnace . . . and to heat them to a white or welding heat, and then shingle them under a forge hammer . . . into half blooms, slabs or other forms. . . . My new invention is to put them again into the same or another air furnace . . . from which I take the half blooms and draw them under the forge hammer . . . into anthonies, barrs, half flats, small square tilted rods for wire or such uses as may be required, and the slabs having been shingled in the foregoing part of the process to the sizes of the grooves in my rollers through which they are intended to be passed, and are worked by me through the grooved rollers in the manner which I use bar or wrought iron fagotted and heated to a welding heat for that purpose. The whole of which discovery or attainment are produced by a more effectual application of fire and machinery . . . than was before known of or used by others and are entirely new and contrary to all received opinions amongst persons conversant in the manufactory of iron.'

Henry Cort's methods, based as they were on existing techniques, subsequently raised much controversy on the subject of priority, not only in Cort's times, but for many years after. Cort's achievement was to gather the existing knowledge and skills, accept the good, reject the bad. He devised a method of iron manufacture and working which opened up a new era in iron manufacture of immediate impact at a time of rapidly expanding industry in Britain. Before his improvements, a tilt hammer, working by water power, produced 1 ton of bars, of doubtful quality, in twelve hours; Cort's rolling mill, absorbing approximately the same power, produced 15 tons of uniformly high quality iron in the same time. The iron, being produced entirely by pit coal, eliminated the problem of timber conservation and, using raw materials of native origin, enabled Britain to become, within a decade, the premier iron-producing country in the world.

Cort negotiated with the ironmasters of Merthyr Tydfil and elsewhere for them to use his patent on a royalty basis and, if things had worked out right, would have been rewarded in proportion to the importance of his innovation. Fate, however, intervened through the death of his partner's father, Adam Jellicoe, and the disclosure of irregularities in his Navy accounts amounting to about £40,000. At this time, Henry Cort owed Adam Jellicoe £27,000 representing much of the capital tied up in the experimental work and the foundry plant. For this, he had given, as collateral security, the rights to his patents of 1783 and 1784; these were immediately appropriated by the Crown Offices, tucked away in the office of the Solicitor to the Crown, where they remained for ten years. In the meantime, Henry Cort, his wife and ten children were left without means of existence, Cort appealing to various officials for some opportunity to repay the debt out of the royalties from the vast output of iron coming from the great ironworks of South Wales and elsewhere. In 1791, the Treasury granted Cort a net pension of £160 per year, the patents were worked free by the ironmasters until they expired in 1797–8 and Henry Cort died on 23 May 1800, a broken-hearted man. Cort was the victim of an era of corruption which existed in Britain at that time. Vast fortunes were being made by peculation of Government funds and by other irregular means, so that the unfortunate inventor, being caught up in this web of financial chicanery, had little hope of redress against enemies unknown to him but who were, nevertheless, responsible for one of the most shameful episodes in Britain's industrial history.

The processes introduced by Henry Cort provided the means

of making iron in the quantities needed by the great industrial expansion of the nineteenth century and, until the invention of the Bessemer process, Cort's puddling methods remained the basis of the manufacture of malleable iron, one of the major civil engineering materials. In 1884, although steel had started to take its place, 8¾ million tons of wrought iron were made in puddling furnaces throughout the world, 2¾ million tons of this in Great Britain, where over 4,500 furnaces were working. Although steel has largely taken its place, wrought iron is still in demand, especially for structural work where corrosion is a problem, and it is made by the processes introduced by Henry Cort, whose little mill at Fontley has become a farm, with no trace of the ironworks except for the mill race which led to his power supply, and hidden dumps of the waste from his furnaces.

The rise of Merthyr Tydfil was the most outstanding example of the impact of Henry Cort's inventions in the hands of ruthless men of industry and enterprise. This town, which lies in the centre of a district supplied by nature with all the requirements of iron manufacture, iron ore, fuel and limestone, was a village of no consequence until the 1750s, when Anthony Bacon leased a district 8 miles long and 4 miles wide for the modest rent of £200 per annum over a period of ninety-nine years. This lease consisted of the whole of the mineral rights of the area, with a small portion of the surface area, sufficient for the erection of a works. On this site, Bacon built an ironworks which he used for manufacturing cannon to be used in the American War and from which he profited greatly until the end of the war terminated his contracts. His rival was John Guest of Dowlais.

About the year 1783, Bacon split up his rights into sub-leases, which he let, for the most part to Richard Crawshay and the remainder to Richard Hill. Guest was already established at Dowlais and the Homfrays, brought to Merthyr by Anthony Bacon, had fallen out with him, establishing themselves at Penydarren. This, then, was the formidable array of enterprise which in 1789 took advantage of the discoveries of Henry Cort and made its leaders the Iron Kings of Merthyr.

Richard Crawshay, the greatest, started from nothing when, in

This conjectural model *in the Science Museum, London, was based on Trevithick's drawing for the Penydarren locomotive.*

1757, he quarrelled with his father and left his home in Norman-ton, Yorkshire, for London. There he was employed in an iron warehouse, married his master's daughter and became the owner. A prize of £1,500, won in a State lottery sent him off to Merthyr, where he bought out the Cyfarthfa interests of Anthony Bacon. He visited Cort at Gosport in 1789 and this led to the adoption of Cort's methods in the Cyfarthfa works. The opening of the Glamorgan Canal in 1795 provided the means of transport which connected the works to the world outside. By 1803, the works were turning out 60 to 70 tons of bar iron weekly and had become the largest in the kingdom. The year 1806 showed even greater expansion with six furnaces and two rolling mills employing 1,500 men at a wage bill of £6,000 per month.

Crawshay's resources about this time may be assessed by the anecdote of his banker friend, Wilkins of Brecon, who, to meet an unexpected call on his funds, appealed to Crawshay who advanced him £50,000 immediately with the promise of another £50,000 should he need it. Richard Crawshay died in 1810 and the works reached its peak under the management of his grandson, William Crawshay the younger.

The Penydarren works of the Homfrays were started in 1784 and although they did not attain the size of the Cyfarthfa or Dowlais works, the name of Penydarren will ever be remembered in conjunction with railways, for it was there that Richard Trevithick, the 'Cornish Giant', made the first steam locomotive ever to run on rails. The Penydarren works was connected to the Glamorganshire canal at Abercynon by a plate tram road built about 1795 by a local engineer named George Overton. Trevithick visited Wales at the turn of the nineteenth century to interest the industrial magnates of the coal and iron industries in his inventions. Homfray became an enthusiastic supporter of Trevithick's ideas and Anthony Hill an equally vigorous opponent. Homfray wagered Hill 500 guineas that Trevithick's machine would haul ten tons of iron from Penydarren to Abercynon Wharf. The first trial took place on 14 or 15 February 1804, and on the 21st of the same month Trevithick's locomotive towed ten tons of iron, five wagons and seventy men the whole distance of nearly ten miles. The journey took just over four hours but included time taken in removing obstructions *en route*. The speed was about 5 miles per hour and, according to Trevithick's account, a boilerful of water with 2 cwt. of coals sufficed.

Not only was Penydarren famous for the Trevithick episode, but in September 1830, the first rail rolled in Wales was produced at the works for the Liverpool and Manchester Railway.

Although the ironworks of Merthyr all prospered during the heyday of the industry, none made such spectacular progress as that of the Crawshays. Under William the younger, the works in 1819 produced 11,000 tons of pig iron and 12,000 tons of bars, and in 1821 the establishment exceeded the whole national output for the ten years between 1740 and 1750. Still prospering,

The wrought-iron links
*of the suspension chains for
Telford's Menai bridge were
subjected to exhaustive tests
before installation.
They were renewed
in high-tensile steel
about 1940.*

Cyfarthfa had eleven furnaces in 1845 which made 45,760 tons of iron and this figure was doubled before the development by Bessemer of mass produced steel brought the iron industry of Merthyr into a slow decline, lasting until 1910 when the great works was closed.

By the beginning of the nineteenth century, both cast and wrought iron had become accepted as materials for major works of construction. Their use, however, depended largely on experience and, until the introduction of large span bridges, this was no serious disadvantage. Thomas Telford, for his design of the Menai and Conway bridges, carried out tests of the materials to be used, especially of the wrought iron for the links of the suspension chains. The development of systematic testing for materials of construction owes most to William Fairbairn, who with the assistance of Eaton Hodgkinson developed the testing of materials to a science, substantially as we know it today.

287

Systematic experimenting *with construction materials owes most to William Fairbairn who developed testing to the science substantially as we know it today.*

William Fairbairn was born in 1789 at Kelso, the son of a farmer, or smallholder, in poor circumstances. He went to the local school, but had to help on the farm. At fifteen he became apprenticed as a mechanical engineer to the Percy Main Colliery near North Shields and on the completion of his apprenticeship moved to London and eventually to Manchester in 1814, still working as a mechanic, but studying in his spare time, as he had done throughout his apprenticeship and after. In 1817, he entered into partnership with James Lillie in opening a business as mechanical engineers and his success in improving the machinery of cotton mills soon brought him into high repute so that by 1824, he was advising manufacturers as far away as Alsace and Switzerland.

The production and use of iron had, by this time, reached such magnitude that Fairbairn, with his interest lying in its economical use, commenced to study its mechanical properties. For this purpose he built an improved testing machine at his works, which became known as Fairbairn's Lever. At this time, he met Eaton Hodgkinson, a mathematician and engineer, whose interest lay in the strength of materials, and invited him to co-operate in the use of the machine. Their first work was done on the properties of cast iron, and included not only tension, compression and bending tests, but also investigations into the effects of temperature and creep. Eaton Hodgkinson, the partner of Fairbairn on many of his researches, and the mathematical part of the team, was born at Northwich, Cheshire, in the same year, 1789, also the son of a farmer, who died while Hodgkinson was a child, so that after an elementary education, he went to work on his mother's farm. The family moved to Manchester in 1811 and there, John Dalton, the scientist, came to know the young man, whose intelligence led to Dalton teaching him mathematics, including the work of the Bernoullis and Euler, which interested him in the strength of materials, from that time to become his life's work. This led to his publication of many important papers on the subject, his election to the Royal Society and, in 1847, his appointment as professor of the mechanical principles of engineering at University College, London.

In the same year, Hodgkinson was appointed a member of a commission to enquire into the application of iron to railway structures. This important enquiry arose from the over-enthusiastic use of iron by some of the early railway engineers, which had

Fairbairn's method of proving a cast-iron girder was by direct loading from a point in the middle using a dead weight on screw gear (below). The structure (right) was tested by repeated impact from the stamper S.

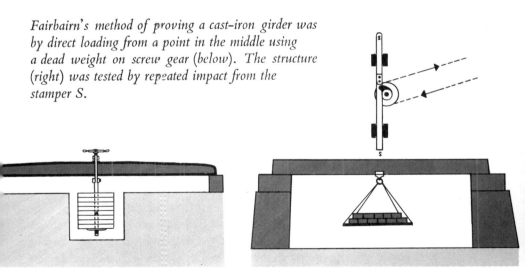

led to a number of failures. The report of the Commissioners includes evidence given by such eminent engineers as J. U. Rastrick, Robert Stephenson, William Fairbairn and I. K. Brunel; it thus provides an overall view of the development of the use of iron immediately preceding the mass production of steel. The first witness was John Urpeth Rastrick, the designer of Chepstow Bridge and a former partner of Hazeldine at Bridgnorth Foundry. Rastrick probably possessed more empirical knowledge of cast iron than any other engineer of his time and it appears that his breaking with the Bridgnorth Foundry arose from his custom of keeping his technical data in a personal memorandum book which he refused to make available to his partner. Much of his evidence was about the quality of iron from various sources. William Fairbairn gave evidence on his method of testing beams of cast iron, which he proved by direct loading using dead weights on screw gear. He also discussed the effect of cross beams resting on the bottom flanges of the main girders of bridges, and concluded that, unless balanced by an equal load on the opposite flange, such loading must lead to harmful torsional stresses. Robert Stephenson's evidence included details of the constitution of the iron used for his high level bridge at Newcastle, information on his collaboration with Eaton Hodgkinson on matters of strength and comments on the effects of impact and fatigue, the latter being, in his opinion, unimportant. Brunel was against the use of cast iron for bridges carrying railways, although he used the material extensively for carrying roads over the railway. For large bridges in cast iron he expressed his preference for arch construction where he could obtain suitable abutments, but indicated his partiality for wrought-iron fabrications, expressing confidence in good riveted or bolted connections under conditions of vibration. Brunel, probably for the first time, stressed the importance of the frictional grip between members riveted or bolted together, a view which was very much in advance of ideas of his time.

The Fairbairn-Hodgkinson combination probably reached its greatest peak of achievement in the experiments leading to the final design of the tubular bridges carrying the Chester and Holyhead Railway over the Conway and the Menai Straits. Robert

Stephenson, who was engineer to the railway, proposed to construct a suspension bridge having tubular girders supported by chains. Fairbairn, who was consulted, experimented with tubular iron beams of different cross-sections and discovered that, unlike cast-iron beams, which failed by tension in the convex face, the thin-walled iron tubes failed by collapsing on the concave, or compression face. In his *Account of the Construction of the Britannia and Conway Tubular Bridges*, published in 1849, he says:

'Some curious and interesting phenomena presented themselves in the experiments—many of them are anomalous to our preconceived notions of the strength of materials, and totally different to anything yet exhibited in any previous research. It has invariably been observed, that in almost every experiment the tubes gave evidence of weakness in their powers of resistance on the top side, to the forces tending to crush them.'

These results, from the first experiments on unstable thin-walled structures, led Fairbairn to pass the problem to Hodgkinson who, as the 'mathematician' of the team was the only one capable of the analysis. Hodgkinson proposed a programme of fundamental tests which, although he made some of them later, were impossible at the time as the construction of the bridge was urgent. Fairbairn therefore built a model in wrought iron, approximately one-sixth full size, having a span of 75 feet. This model was provided with a cellular structure on the compression or upper face and was tested to failure. The first failure was on the bottom side, which was strengthened and further tests to failure were performed, reinforcing until the top, bottom, and sides of the tube were equally stable. The design of the bridge then proceeded on the assumption that the carrying capacity of the tube increases as the square of the linear dimensions and that its weight increases as the cube. This great experiment not only included the effects of dead weight on the structure, but also a study of the effects of wind pressure and of sunlight. It undoubtedly initiated a new approach to the problems arising from the design of large structures, and their study by model, which has continued with advantage to the present day.

The Britannia tubular bridge *was the result of a series of tests to destruction on large models with progressive improvements after each test. The value of*

such methodical testing is proved by the fact that trains of modern weight are still able to use the bridge on their regular runs.

Bessemer's converter forced air through the molten metal, removing the carbon and most of the impurities.

The first half of the nineteenth century was the period in which iron reached its highest peak of demand. The construction of railways had called for enormous quantities and mechanical inventions of all kinds absorbed a large tonnage. The cost of malleable iron was, however, affected by the laborious procedure required by the puddling process, which no subsequent work had been able to improve on the methods introduced by Henry Cort. In the 1850s, however, the call for armaments for the Crimean War aroused the interest of an engineering and metallurgical inventor, Henry Bessemer, who was born on 19 January 1813, the son of a retired engineer living in Hitchen in Hertfordshire. Bessemer, from his earliest days, had been interested in mechanical things, especially those connected with the use of metals and of metallurgy. He first experimented with a reverberatory furnace, passing a blast of hot air over the surface of the molten iron and thereby succeeded in producing a malleable iron of low carbon content. He realized before long that air alone, if brought into contact with a sufficiently extensive surface of molten pig iron, would rapidly convert it to a low carbon iron. Pursuing this idea, he experimented with a series of furnaces designed to allow a blast of air to be forced through the molten metal, thus burning out the carbon and most of the impurities. By 1856, Bessemer had constructed a converter with a capacity of 7 cwt. of molten pig iron in his workshops at Baxter House, together with a cupola for melting the raw iron. He had a steam driven blower for the air, and an ingot mould with hydraulic ram ejection. His first blow of 7 cwt., produced a pure, homogeneous ingot in about thirty minutes' blowing, with little labour and no fuel for the converter. On the ingot cooling, he carried out his first test, using the only implement available, a sharp axe, which he used to cut into the ingot to prove the malleable nature of the material.

Bessemer invited George Rennie to view his process and, following the blow, Rennie was so impressed that he persuaded Bessemer to read a paper on his invention to the summer meeting of the British Association to be held in Cheltenham the following week. George Rennie was President of the Mechanical Section that year and, by his influence, Bessemer's paper was the first on

The furnace *on the left is at the 'blowing' or refining stage. The other pours the refined metal into the ladle which revolves to fill the semicircle of ingot moulds. This painting is of a Bessemer steel plant at Ebbw Vale in about 1860.*

the agenda. It was presented at the meeting on the morning of 13 August 1856, possibly the most important paper ever presented to the British Association and certainly the most important event in the history of Cheltenham. Bessemer's paper announced a revolution in the technology of iron and its impact on the industry was immediate, but the inventor was not entirely free of trouble, as later trials proved that his converter was only capable of making satisfactory steel from ores free of phosphorus and sulphur. The cure for this was to come later, but meanwhile good quality steel could be produced in quantity from ore free of the troublesome elements and at prices far below those of puddled iron.

Bessemer's process had been established for about twelve years when two separate attempts to make steel in the open hearth furnace both matured at about the same time. Dr C. William Siemens, a famous man of science, in 1866 obtained a British patent for producing steel in the open hearth while, in France, Pierre and Emile Martin succeeded in the same object. The open hearth furnace offered advantages over the Bessemer converter; it would melt any proportion of scrap in the mix, the furnace could be built to take much larger tonnages and, although the process was slower, this gave the advantage that better control of the final product was possible. The problem of phosphorus and sulphur remained, and was first met by a discovery of Mushet who fed a small quantity of ferro-manganese or spiegeleisen into the molten steel just before tapping; this had a purifying effect in reducing the unwanted elements while increasing the percentage of manganese, a beneficial element.

The real answer was found by a clerk, Sidney Gilchrist Thomas and his cousin, Percy Carlisle Gilchrist, a metallurgist, who in 1879, announced to a meeting of the Iron and Steel Institute that if, instead of the customary acid furnace lining of ganister or silica sand, a lining of a basic mineral such as magnesia or dolomite is used with slag of a basic material such as limestone, the phosphorus would be absorbed in the basic slag and thus entirely eliminated; the sulphur similarly would be materially reduced. Thus, vast resources of iron ore, previously unusable, became available for the manufacture of steel by what became known as the 'basic' process.

The mass production of uniformly reliable steel at a low price had an almost immediate and profound effect on civil engineering. The railways, in particular, achieved an immediate benefit from the production of good rails for the permanent way. The importance of steel for this purpose may be assessed from figures showing the growth of railways throughout the world. In 1850, the world mileage of railways was 18,000; in 1860 this had increased to 63,000 miles; by 1870 it was 127,000; and in 1878 the total of running line was 206,000 miles, to which could be added sidings, loops, etc., to make a world mileage of 250,000, requiring about 30 million tons of rails. Such vast quantities could not possibly have been produced by the puddling process, nor

The age of steel *was heralded by the introduction of the Bessemer
converter and the blast furnace. Royalty visited a Sheffield factory
in 1875 and watched the casting of steel ingots.*

Modern steel construction *aids the astronomer in his quest for know-ledge of the universe by means of the Jodrell Bank radio telescope.*

could the more frequent renewal of wrought-iron rails have been tolerated on the long lengths of railway coming into use in the great open spaces of the world. It may be accepted, therefore, that if railways originated with wrought-iron rails in the highly populated countries, the opening up to that means of transport of new continents was due to the steel of Bessemer, Siemens and the Martin brothers.

The low cost and uniformity of steel also made it an attractive material for bridges. By 1863, three steel bridges were being built for the Netherland State railways; these were of lattice girder construction, having a span of 30 metres each. The steel for these, in the form of plates, angles and rivets, was supplied at a cost of £250 per ton. In 1864, a steel swing bridge was built by S. B. Worthington of Bessemer plates to carry a railway over the Sankey Canal and, in 1865, the London and North Western Railway constructed a bridge of steel supplied by Henry Bessemer and Co., of Sheffield. The design of steel bridges at that time was based on wrought-iron practice, the steel sections being reduced to five-eighths. Fowler and Baker, in 1864, designed a steel bridge as a continuous girder having spans of 1,000 feet for the South Wales–Great Western Railway link. This span was later reduced to 600 feet, an Act of Parliament obtained and a contract let. Financial difficulties, however, prevented the construction of a bridge at that time. The same engineers were, however, given their chance to do something big when, in 1880, they commenced the designs for the Forth Bridge in which, by the time it was completed in 1890, over 50,000 tons of steel had been used for the superstructure alone.

By the end of the century, construction in steel had become normal practice. Riveting and bolting were the methods of fastening the component parts of the structural whole, and designs were conditioned by the technical needs of these fixings. Early in the twentieth century, however, the use of the electric arc as a means of welding steel became a practicable proposition, thus freeing the designer from the necessity of providing for rivets. This freedom, slow to be appreciated at first, ultimately led to the design of structures having clean, elegant lines, and a beauty comparable with the best of the Greek and Renaissance periods.

Steel has also enabled the civil engineer to make the most of another material of ancient origin, but of modern improvement—concrete. Although in many periods of history, mixtures of stones and mortar had been used in works of construction, they had mostly been intended to fill the voids in masonry or brickwork.

299

For many centuries after the time of the Romans, the secret of their cement lay in abeyance and all mortar was made with lime. The great expansion of industry during the eighteenth century led civil engineers to interest themselves in the properties of cement, and, in particular, John Smeaton made a study of the properties of various mixtures of lime, clay, trass and pozzolana. He described these experiments in his *Narrative of Eddystone Lighthouse*. He concluded that a proportion of clay mixed with limestone, which by burning is converted into a brick, acts more strongly as a cement. He also found that Italian pozzolana from Civita Vecchia, when mixed with Aberthaw lias, made a cement which would harden under water.

On 21 October 1824, a patent was granted to Joseph Aspdin, of Leeds, for 'a superior cement resembling Portland stone'. Joseph Aspdin was born on Christmas Day 1778, in the parish of Hunslet, near Leeds. In his patent application he described himself as a bricklayer and refers to 'my method of making a cement or artificial stone for stuccoing Buildings, Waterworks, Cisterns, or any other purpose to which it may be applicable (and which I call Portland cement) is as follows. . . .' Aspdin's cement was not the same as that which we recognize as Portland cement today, his specification gave no proportions of limestones and clays, neither was it burnt at as high a temperature as today. It was, however, used by Marc Isambard Brunel, in 1828, for the Thames tunnel, in spite of its high cost—about 20s. to 22s. per barrel as against Roman at 12s. per barrel. John Grant, the engineer for the London Main Drainage scheme in 1859, used Portland cement extensively for the works and safeguarded the scheme by specifying tests of the cement. This testing was done at the site on machines built specially for the purpose, and in the six years that construction continued, 70,000 tons of Portland cement were used, and 15,000 tests were made. This must have been the first instance where a series of laboratory tests were carried out at site. Aspdin made his cement in what was called a 'bottle' kiln, burning coke. It was intermittent in working, as the coke and stone were put into the kiln in layers and a charge was

drawn at intervals, followed by a period of cooling to allow the filling to be repeated. It was expensive in fuel and much of the clinker was underburnt.

Modern Portland cement is largely due to the work of I. C. Johnson, who was born at Vauxhall in 1811. After being apprenticed to a building firm, he joined J. B. White and Son of Swanscombe, Kent, becoming their works manager at the age of twenty-four. He started on his own account as a cement manufacturer in 1850 and turned his enterprise into a limited company in 1894. In 1872, he was granted a patent, No. 1583, for 'Improvements in the manufacture of Portland and other cements'. He died when over 100 years of age, on 29 November 1911. Johnson's discovery was the result of a long period of experimenting and testing; at each step he carefully considered his results and from these, planned his next experiments. His discovery has

Rotary kilns, *seen here at the Swanscombe works, facilitated the mass production of cement.*

been attributed to his accidental burning of clinker and it is true that he did this, but the real secret of his success was the knowledge already gained from his previous studies, which led him to powder the clinker and test it as a cement. Johnson's account of his experiments was quoted in the *Building News* of December 1880, and the vital part reads as follows:

'I was at this time (about 1845) manager of the works of Messrs White at Swanscombe, making only the Roman cement, Keene's plaster, and Frost's cement, the latter composed of 2 chalk to 1 of Medway clay, calcined lightly and weighing 70 to 80 lb. per bushel.

'My employers, attracted by the flourish of trumpets, that was then being made about the new cement, desired to be makers of it, and some steps were taken to join Aspdin in the enterprise, but no agreement could be come to, especially as I advised my employers to leave the matter to me, fully believing that I could work it out.

'As I before said, there were no sources of information to assist me, for although Aspdin had works, there was no possibility of finding out what he was doing, because the place

The influence *of tradition is seen in the simulated joints of masonry voussoirs in the first concrete bridge to be built in England at Seaton in 1877.*

Weaver's Mill, *built at Swansea in 1898, was one of the first reinforced concrete, multi-storey buildings in Britain.*

was closely built in, with walls some 20 feet high and with no way into the works, excepting through the office.

'I am free to confess that if I could have got a clue in that direction I should have taken advantage of such an opportunity, but as I have since learned, and that from one of his later partners, the process was so mystified that anyone might get on the wrong scent—for even the workmen knew nothing, considering that the virtue consisted in something Aspdin did with his own hands.

'Thus he had a kind of tray with several compartments, and in these he had powdered sulphate of copper, powdered limestone, and some other matters. When a layer of washed and dried slurry and the coke had been put into the kiln, he would go in and scatter some handfuls of these powders from time to time as the loading proceeded, so the whole thing was surrounded by mystery.

'What then did I do? I obtained some of the cement that was in common use and, although I had paid some attention to chemistry, I would not trust myself to analyse it, but I took it to the most celebrated analyst of that day in London, and spent some two days with him. What do you think was the principal element, according to him? Sixty per cent of phosphate of lime! All right, thought I, I have it now. I laid all the neighbouring butcher's under contribution of bones, calcined them in the open air, creating a terrible nuisance by the smell, and made no end of mixtures with clay and other matters contained in the analysis, in different proportions and burnt to different degrees, and all without any good result . . .

'I had a laboratory and appliances on the premises, so I worked night and day to find out the component parts of the stones from Harwich and Sheppey. Having found these and having tried many experiments, spreading over some months, in putting different matters together, I began to think that lime and alumina were the chief ingredients necessary. I therefore, tried quicklime powdered and mixed with clay and calcined, by which means I got something nearer. It was a cement very much like Frost's. After this I used chalk and clay as used in Frost's cement, but with more chalk in proportion. The resulting compound being highly burned, swelled and cracked.

'By mere accident, however, some of the burned stuff was clinkered, and, as I thought, useless, for I had heard Colonel Pasley say that he considered an artificial cement should feel quite warm after gauging, on putting your hand on it, and that in his experiments at Chatham he threw away all clinkers formed in the burning.

'However I pulverised some of the clinker and gauged it. It did not seem as though it would harden at all, and no warmth was produced. I then made mixtures of the powdered clinker, and powdered lightly-burned stuff, this did set, and soon became hard. On examining some days later the clinker only, I found it much harder than the mixture; moreover, the colour was of a nice grey.

'Supposing that I had nearly got hold of the right clue, I proceeded to operate on a larger scale, making my mixture of 5 of white chalk to 1 of Medway clay. This was well burned in considerable quantities, and was ground finely; but it was, of course, a failure from excess of lime, although I did not then

Pre-cast concrete *was used by Pier Luigi Nervi for Rome's Flaminio Stadium.*

know the reason of it. The whole of this material was tossed away as useless into a kind of tunnel near at hand, and laid there for some months, after which I had the curiosity to take

305

a sample of it and gauged it as before, when, to my astonishment, it gauged smoothly and pleasantly, and did not crack and blow as before, but became solid, and increased in hardness with time.

'Cogitating as to the cause of this difference, it occurred to me that there had been an excess of lime, and that this exposure in a rather damp place had caused the lime to slake.

'This was another step in advance, giving me as it did, the idea of there being too much chalk, so I went on making different mixtures until I came to 5 of chalk and 2 of Medway clay, and this gave a result so satisfactory that hundreds of tons of cement so mixed were soon afterwards made.'

The mass production of cement could not be achieved by the intermittent processes using kilns based on the 'bottle' form. The turning-point in this respect came with the invention by Crampton, in 1877, of the rotary kiln. This important innovation was not an immediate success, and several variants were patented, the first actually being built at Arlesley, near Hitchin, in 1887.

This kiln consisted of an inclined cylindrical furnace supported on roller bearings and turned through worm gearing; the firebrick lining was ridged to agitate the contents, which were fed into the upper end from a hopper to travel slowly down, while being slowly heated by a producer gas flame, to fall from the lower end into a pit.

The first *rotary kiln actually to be built was at Arlesley in 1887. This was 25 ft long and 5 ft in diameter.*

VIII

Greek and Roman principles — Leonardo da Vinci
Galileo and the science of dynamics — mathematics
and the calculus — Newton's laws of motion
Hooke's Law — the Bernoullis — Monge and descriptive
geometry — Young's Modulus — soil mechanics
Navier's work on materials — 'Theorem of Three
Moments' — Maxwell and Mohr — Castigliano and
the principle of least work — modern developments

◄One of the founders *of modern science, Galileo's work ranged over almost the whole of physics and included the invention of the thermometer.*

Structural Theory

UNTIL RECENT TIMES, structural theory, based on combined mathematical and experimental investigation, was practically unknown. The architects and master masons of earlier days possessed a great body of empirical knowledge which was based on experience, and much of which was kept secret from all but those dedicated to the profession of building. That they had such knowledge is demonstrated by the remaining relics of buildings by the Egyptians, Greeks and Romans, but little of this secret lore survived the long, dark period which followed the fall of Rome, so that the Renaissance meant a start almost from the beginning. A little of the written records did, however, survive, including some of the work of Archimedes, the Greek mathematician who lived from 287 to 212 B.C. He combined theory and practice in a number of mechanical and structural applications, advancing the knowledge of statics, which already existed, by giving a rigorous proof of the principles of the lever and laid down the basis of knowledge of the centres of gravity of bodies.

Another ancient writer, whose work has survived, was Vitruvius, whose book *De Architectura* described the building methods of his time. A great engineer and architect, he lived in the time of the emperor Augustus, at the peak period of construction of great works of engineering, in which he took a prominent part. Without his writings, little would be known of the principles on which the Greek and Roman engineers executed their works of construction and the writings of other authors, using technical terms current in Roman times, would have been difficult or impossible to understand without reference to the full descriptions given by Vitruvius. In the eleventh chapter of his book six, Vitruvius informs his readers that:

'when lintels or beams are loaded, they are apt to sag in the middle, and cause fracture to the work above; but when posts are introduced, properly wedged up, this is prevented: by the insertion of two inclined pieces of timber, it may also be accomplished. The weight of the wall may be discharged by arches formed of wedges, concentrically arranged; these, turned over the beams or lintels, relieve the weight, and prevent them from sagging. In all buildings where piers and arches are used, the outer piers are to be made wider than the others, that they may resist the thrust of the arches.'

On no other theory but such empirical rules, the Romans built the great structures of which many still survive after twenty centuries of nature's efforts to destroy them. Wasteful in material and labour though they may appear to us today, the Roman economy, based on the subjugation of a vast empire, was well able to afford to build and maintain them.

The Roman Empire crashed, the written records disappeared, but some remnants of craft knowledge survived during the Dark Ages to enable the mediaeval builders to begin again. The art of construction, through an organized body of itinerant masons, carpenters and metal workers, was developed to a new peak of achievement culminating in the building of the great ecclesiastical structures of the Middle Ages. No works of art of any kind can surpass those of the masons of that period who, without benefit of structural theory, combined function with form in their columns, roofs, steeples, flying buttresses and so on in buildings which still stand to the glory of God and the men who created them.

The artist-builders of the Middle Ages led to the artist-engineers of the Renaissance. The new spirit of enquiry which arose influenced the art of building. The combination of constructive design with the decorative arts in the activities of one individual practitioner had survived from the earlier age, and it was common for great artists of the time to engage themselves in works of architecture and engineering for their princely patrons. Such names as Michelangelo Buonarotti, Benvenuto Cellini, and, above all, Leonardo da Vinci represent the great practitioners of

Intricate fan vaulting *in the roof of King's College Chapel at Cambridge demonstrates a new peak of building achievement in the Middle Ages.*

the age. Leonardo da Vinci, that incredible man, born in 1452, not only left to us some of the greatest pictures ever painted, but a series of notebooks in which he illustrated and described ideas so advanced that only now can modern technology put some of them into practice.

A true representative of the scientific age just beginning, Leonardo based many of his ideas on actual experiments, and his investigations included tests of the strength of materials; the analysis of systems of pulleys and levers; the forces in an arch; the strength of beams and the forces in triangulated structures. He was probably the first man to study the forces in structural frames by the application of the principles of statics and to determine the strength of materials by controlled experiments. Unfortunately, Leonardo wrote all his notes in a mirror script and for many decades his work was unknown and from his death in 1519 the design of structures remained a purely empirical exercise.

The real start of an attack on structural problems was made by

A true representative *of the scientific age just beginning, Leonardo da Vinci based many of his ideas on actual experiments. This model is of his design for a timber truss bridge.*

Galileo Galilei, a member of the Florentine aristocracy, who was born in Pisa in 1564. Educated by the Church and at Pisa University, he soon became interested in mathematics and mechanics, apparently through the influence of Leonardo da Vinci's discoveries. At twenty-five years of age, he became professor of mathematics at Pisa and worked on his experiment with falling bodies which led to his treatise *De Motu Gravum* of 1590—the birth of the science of dynamics. This treatise, and his subsequent writings on astronomy, were so contrary to the accepted teachings of the times that they led Galileo into conflict, first with the supporters of the teachings of Aristotle and, later, with the Church. His condemnation by the Inquisition following the publication of his book on the planetary system in 1632 led to his recantation and his retirement from public life. From 1634 to his death in 1642, he devoted himself entirely to his studies in strict privacy and in 1638, from the press of the Elzevirs of Leiden, appeared his *Discorsi e Demonstrazioni Matematiche*, the

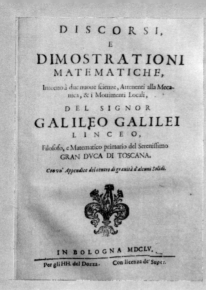

The 'Two New Sciences' of Galileo, first published in 1638, initiated
the study of the strength of materials and dynamics.

'Two New Sciences'. The sciences were the strength of materials
and dynamics, and Galileo's book initiated the study of the
properties of materials of construction and the mechanics of
elastic bodies, although Galileo himself treated his materials as
inelastic. From his conclusions, however, later investigators built
up a vast body of knowledge on elasticity, without which modern
engineering would not be possible.

In his book Galileo discusses, first, the behaviour of materials
in tension, finding that the strength of a bar is proportional to its
cross-sectional area and independent of its length. He uses the
expression 'absolute resistance to fracture' to describe its ultimate
strength and proceeds to investigate the behaviour of the same
bar when used as a loaded cantilever. He found that the 'resistance'
(moment of resistance) of a beam was proportional to its breadth,
the square of its depth, and inversely proportional to its span. He
assumed, however, that all the fibres, except those at the compression edge which formed the fulcrum, were in tension and
equally stressed. Although this assumption is not correct, Galileo

314

was able to use this theory to establish several important points. He states that any given prism of rectangular form will offer greater resistance to fracture when standing on edge than when lying flat and this in the ratio of the width to the thickness. He also concluded that the bending moment, due to the weight of a beam, is increased as the square of the length. He found that, with geometrically similar cantilever beams under the action of their own weight, the bending moment at the built-in end increases as the fourth power of the length, but the resisting moment is proportional only to the cube of the linear dimensions; he therefore concluded that the dimensions of a structure must ultimately be limited by the weight of its own materials of construction. Galileo studied the properties of cylinders and hollow beams and found that, although the absolute strengths of hollow and solid sections of the same area are the same, their moments of resistance are equal to their absolute strength multiplied by the outer radius, and that therefore the bending strength of a tube is proportional to its diameter.

René Descartes

Structural theory, for its development, is dependent on mathematics as a working tool of primary importance. No history of the subject could, therefore, be fairly presented without some reference, in passing, to those mathematicians whose contributions have made the theory of structures possible. One of these, René Descartes, was born when Galileo was thirty-two years of age, in 1596 at La Haye, near Tours in France. The descendant of an ancient family in moderately affluent circumstances, Descartes was delicate in his youth and his formal education did not begin until he was eight, when he was sent to a Jesuit college. There he was permitted, because of his health, to lie in bed every morning, a practice which remained with him all his life, and to which he attributed much of his ability in philosophy and mathematics. Descartes lived in the environment and period described by Dumas in *The Three Musketeers* and, at eighteen years of age, left home and began a career of intermittent soldiering of the swashbuckling kind which lasted until he was thirty-two, when he retired to Holland where he remained for twenty years, never settling down, but all the time developing his ideas. In 1637,

Almost together *Isaac Newton and Gottfried Leibniz (right) arrived independently at the calculus, the most important tool in the mathematical equipment of the engineer.*

against his own reluctance, his friends persuaded him to publish his great work, known shortly as *The Method*. In this, he introduces to the world his system of co-ordinate geometry, the greatest gift ever given to the artist-engineer whose mind, receptive of visual images of ideas, often tends to rebel at the (to him) more arid approach of the analytically minded. Descartes provided the link which enabled the two to come closer together, by his method of plotting the values of an equation on a system of co-ordinate lines, he enabled the visually minded to *see* what was happening and to grasp firmly the implications of the mathematics.

Descartes died in 1650, leaving co-ordinate geometry as his greatest gift to science, eight years after the birth in 1642 of Isaac Newton at Woolsthorpe, in Lincolnshire, and four years after that of Gottfried Wilhelm Leibniz in Leipzig. The first of these was probably the great mathematician of all time, the second,

not far behind; and, almost simultaneously, each produced the most important tool in the mathematical equipment of the engineer—the calculus. Newton, like Descartes, was delicate in his youth and was educated at the village school, Grantham Grammar School and Trinity College, Cambridge. His studies, influenced by Galileo, Kepler and Descartes, led him to state his three laws of motion, on which the whole science of dynamics rests. These laws, in turn, require a special mathematics for their full development especially for operations connected with the second law which states: 'Rate of change of momentum is proportional to the impressed force and takes place in the line in which the force acts.' The solution to the problem of measuring those rates of change was found by Newton in his differential calculus. A further problem arising from the first was the calculation of the total effect in a given time of a variable which is changing from instant to instant; this was solved by the development of the integral calculus and Newton finally crowned his masterpiece by the discovery of the intimate and reciprocal relationship of the two.

Newton, like many other discoverers, was loath to publish his work, and it therefore remained unpublished for some years. The actual dates of his discoveries in the calculus, universal gravitation and the nature of light are not known, but all were achieved while the university was closed for two years in 1665 and 1666, the years of the Great Plague (bubonic plague). Leibniz, at this time, was a law student at Leipzig with no intention of adopting mathematics as a career; in 1672, however, he met Christian Huygens, who interested him in physics and, by a natural transition, in his true vocation, mathematics. By 1675, Leibniz had worked out the basic concepts of calculus and in 1677 published the work, following which Newton published his. The simultaneous publication of this monumental discovery led to an acrimonious argument on priority in which, at first, the two discoverers did not take part; but eventually, when the verbal and written war reached international proportions, they also were drawn in. As a result of this unfortunate squabble, English mathematicians for a century ignored the work being done on the Continent in developing the calculus as a tool for science, to the great loss of engineering and technology.

If Newton said: 'If I have seen a little farther than others, it is because I have stood on the shoulders of giants', his meaning may be interpreted as expressing the inevitability of the ultimate discovery, by someone, of the calculus and his other major achievements. The work of Galileo, Descartes, Kepler and others had advanced the mathematical sciences so far that the simultaneous discovery by Newton and Leibniz of the calculus was no coincidence; the measure of Newton's genius was, however, the immense advances he made in science during the two years spent at Woolsthorpe in 1665 and 1666 as a refugee from the Great Plague. Certainly the rate of change in scientific discovery, including structural theory, took an upward step as a result of the calculus.

Seven years older than Newton, and in his youth, another

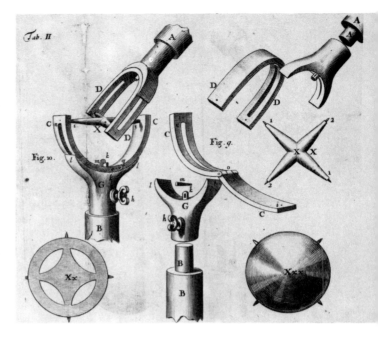

First curator of *experiments of the Royal Society, Robert Hooke formulated the law that bears his name. His inventions included the universal joint which today has a multitude of applications.*

delicate child, was Robert Hooke, born in 1635, the son of a vicar in the Isle of Wight. While still very young, he showed an interest in things mechanical, as had Newton and, after schooling at Westminster, he became a chorister at Christ Church, Oxford, taking his M.A. degree in 1662. In this year, the Royal Society received its first charter, and on the recommendation of Robert Boyle, Hooke became its first curator of experiments. During this period Hooke worked on problems in optics and made important discoveries including the causes of interference colours of soap bubbles and the phenomenon of Newton's rings. In 1678, he produced the first published paper discussing the elastic properties of materials *De Potentia Restituva* ('Of Spring'). This paper, first delivered as a lecture, included the famous law which now bears his name *Ut tensio sic vis* (as the pull, so the stretch). He first discovered this in 1660, but kept it secret until 1676 when he published it in anagram form *ceiiinossstuu* doubtless to secure his priority. Hooke's other inventions included the well-known universal joint, which today has a multitude of applications.

319

Hooke's Law now forms the basis of the mathematical theory of elasticity and, as with other discoveries of importance, was arrived at independently and almost simultaneously by another investigator, E. Mariotte, prior of St Martin-Sous-Beaune, at Dijon, who in 1680 produced the same law based on experimental work, the methods for which he introduced into French science through the French Academy of Sciences. Mariotte's other discoveries included the ballistic pendulum, the laws of impact and, simultaneously with Robert Boyle, the law of gas pressure which states that, at a constant temperature, the pressure of a fixed mass of gas multiplied by its volume remains constant.

Mariotte developed his investigations into elasticity much further than Hooke, mainly as a result of his work on the design of the pipe lines for the supply of water to the Palace of Versailles. Through this he became interested in the strength of beams and by experiment found that Galileo's theory of bending gave unduly high figures for the breaking load and, for the first time, took account of the elastic properties of the materials he was testing. Unlike Galileo, who assumed that all fibres in a loaded cantilever were in tension, Mariotte noted that the fibres in the lower half of the section at the point of support were in compression. His experiments with beams led him to announce that a beam with built-in ends would carry twice the central load of a simply supported beam of similar dimensions and his work on pipes under water pressure led him to arrive at a formula for their bursting strength.

The seventeenth century started a great spate of writings on mathematics and the mechanical arts which survived all the vicissitudes of wars and revolutions of the time. In 1609, Holland became an independent country and it is significant that many of the scientific writings of the time were printed in Leyden and Amsterdam. In 1649, Charles I was executed, six years after King Louis XIV began to rule France for seventy-two years. In 1660, Charles II became King of England after his French exile; the influence of the French Academy of Sciences had impressed him and his support led to the formation of the Royal Society in England. The publication of works and the interchange of ideas sponsored by these two scientific societies led ultimately to the very great expansion of scientific knowledge in the eighteenth and nineteenth centuries, an expansion which, in the present century, is still increasing.

The unsettled times of the sixteenth century drove many people into exile. England benefited by many of these immigrants, who because of political or religious intolerance, brought their native arts and crafts with them, as, for instance, the silk and paper industries. One such family settled in Basle, in Switzerland. They were the Bernoullis, probably the most prolific and continuous line of mathematicians of all time. Their most fruitful phase is given by the genealogical tree below, but even later generations still consistently produced brilliant men, although not necessarily mathematicians:

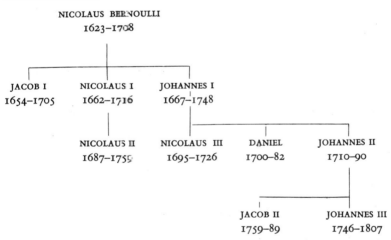

Of these nine descendants of the original Nicolaus Bernoulli (1623–1708), no less than eight became eminent mathematicians, the strain lasting for more than a century and for most of this time, i.e. from 1699 to 1790, there were Bernoullis on the list of foreign members of the French Academy of Sciences. Greatly stimulated by the work of Leibniz, the Bernoullis found in the calculus the ideal tool for solving problems in mechanics and physics. Jacob I made importance advances in the theory of bending of beams; he originated the concept of a beam consisting of

Daniel Bernoulli's work
*on mathematical physics and
hydrodynamics is known to
all civil engineers.*

filaments capable of being stretched or compressed and the effect
of the resistance arising therefrom. In particular, his work on
deflection curves led others to examine the problem further, and,
more successfully, Johannes I formulated the principle of virtual
displacements. To engineers, however, Daniel was the greatest of
them all. At eleven years of age, his brother Nicolaus started
teaching him mathematics; in 1725, at twenty-five years of age,
he became professor of mathematics at St Petersburg, where he
remained for eight years before returning to Basle where he
ultimately became professor of physics. He worked on a wide
range of mathematical physics and laid the foundation of that
subject; his work on hydrodynamics is known to all civil engineers
while, in the elastic properties of materials, his own inspiration
was combined with one even greater, that of his friend and pupil
Leonard Euler.

Euler, described as the most prolific mathematician in history,
was the son of a Calvinistic minister, and was born in Basle on
15 April 1707, where the Bernoullis were already established.

Paul Euler, the father, had been a pupil of Jacob I and taught his son mathematics at an early age. Leonard soon came directly under the Bernoulli influence and, from then on, his life became dedicated to a mathematical career. At twenty years of age, he followed his friends Daniel and Nicolaus III to the Academy at St Petersburg where, in 1733, he followed Daniel into the chair of mathematics. After seven years there, he went to the Berlin Academy for twenty-four years, then back to St Petersburg for the rest of his days. Each of these moves brought him royal patronage, a valuable asset in those days. At St Petersburg, he came under the interest of Catherine the Great of Russia, and in Berlin, of Frederick the Great of Prussia; by their interest in Euler, these two people probably have greater claim to permanent fame than from all their other activities. Euler's special interest to the civil engineer lies in the very full working out of Jacob Bernoulli's ideas of the elastic line in the bending of beams of constant and variable section, under different conditions of loading, straight or curved initially. He also initiated the subject of elastic stability, particularly as applied to columns, by showing that short columns fail by simple compression, but those of more than an optimum length fail by bending. His thesis on this subject entitled *Sur la Force des Colonnes* was submitted to the Berlin Academy in 1759 and a further series of papers to the Academy of St Petersburg in 1778 gives the relationship between the load, the stiffness of the column, and its length, showing that the stiffness has the dimensions of a 'moment of inertia', being expressible as the product of a force and the square of a length. One only of Euler's contributions to civil engineers, his column formula, would have entitled him to fame; but this was merely one special result of his greater work on deflection curves, in itself a small portion of Euler's life output of useful mathematics.

Leonard Euler

The growth of communications following the Renaissance led to a great impetus to the art of bridge building. It was therefore inevitable that before long the purely empirical methods of the Romans began to give way to developments based on theories of the behaviour of arches. In 1695, Philippe de la Hire, a member

323

of the French Academy, published a short paper on the analysis of the arch. In this paper he compares the tensions in a loaded chain or cord with the thrusts in an arch, an idea already suggested by Robert Hooke, and for the first time uses geometry to assess the forces in what we now know as the funicular or link polygon. La Hire's semicircular arch is assumed to have perfectly smooth faced or frictionless voussoirs and demonstrates that such an arch has no stability; he therefore points out that the cement must contribute to the stability. Not all engineers of the eighteenth century accepted La Hire's arch theories. Hubert Gautier, in his *Traité des Ponts*, published in 1716, described experiments on a small model built with wooden voussoirs, and strongly criticized La Hire's theory. La Hire's methods were, however, used extensively by the French eighteenth-century engineers in their work on arches and his smooth voussoir theory survived until the end of the eighteenth century. His link polygon, was, however, his greatest and most permanent contribution to the equipment of the engineering designer.

The greatest contributor in the eighteenth century to structural theory, and the one of most permanent influence, was Charles Augustin Coulomb, who was born at Angoulême in 1736, and after being educated in Paris, became a military engineer. Coulomb was another great innovator whose most productive period occurred through circumstances which gave him the opportunity to think and work without undue interference; he was posted to Martinique, where he served for nine years. While on the island, he wrote his paper on *The Application of the Rules of Maxima and Minima to Statical Problems relating to Architecture*. In this, he presented his ideas as a series of propositions, corollaries and remarks; he dealt with the laws of equilibrium, the resolution of forces, friction, cohesion and the behaviour of beams. For the first time, he recognized the importance of compression as well as tension in a beam under load and he also demonstrated the existence of a vertical component, or shearing force. In this monumental essay, Coulomb clearly stated that:

(1) The sum of the tensions must balance the sum of the compressions.

(2) The sum of the vertical components of these internal forces must equal the load applied. (The first clear recognition of shearing force.)

(3) That the sum of the moments of the internal forces (*the moment of resistance*) must balance the *bending moment* produced by the loading.

For the first time, the equilibrium of forces acting on a beam was clearly stated; the advance of knowledge was great, and the effect of Coulomb's work has been permanent.

The same paper introduced a new method of determining the conditions of stability of an arch, based on the knowledge gained by that time—from experiments—that arches rarely fail by the causes implied by La Hire's smooth voussoir theory, but rather that they collapse by the relative rotation of their segments due to moments arising from the horizontal and vertical thrusts. Unfortunately, Coulomb did not develop his arch theory to suit the needs, nor the understanding, of practising engineers of the time, and until the nineteenth century his work on arches was not fully appreciated.

Coulomb also contributed a notable addition to the theory of earth pressure, the concept of the wedge of pressure which, within certain limits and applied to cohesionless soils may still be applied today. He carried out experiments in torsion and for this purpose used a torsion balance of considerable accuracy which enabled him to describe correctly the behaviour of a material subject to torsion. The torsion balance was developed by him for his researches in the measurement of small electrical and magnetic forces for which work his name is commemorated in the electrical unit of quantity. His writings on many other scientific subjects were of equal importance, and it may fairly be said that his work provided as important a turning-point in structural theory as it did in many other branches of science. He died in 1806, a new century which was to show, by the development of his ideas, its appreciation of his work in science.

In all scientific work, and not the least in engineering, ideas are put forward and theories expounded through many channels; it is thus necessary at intervals for someone to review progress, sift out the sound and essential knowledge and put it into an organized form. Of these benefactors to engineers, one of the most important was Bernard Forest Belidor, born in Catalonia in 1693 and found as an orphan child, five months old in enemy country, by an artillery officer who adopted him. He became a military engineer and a professor at the School of Artillery at La Fère. Belidor found a lack of textbooks in the technical schools of his day and, to fill the gap wrote his own, *La Science des Ingénieurs* in 1729, an elementary book on mathematics in 1735, and *Architecture Hydraulique* in 1737. His engineering books enjoyed a great sale and the last edition of his engineering science was published 101 years after the first. Another technical author of the same period was Hubert Gautier (1660–1737) who, after training as a doctor of medicine, turned to engineering, and served for twenty-eight years as engineer to the province of Languedoc, becoming in 1716 *inspecteur des ponts et chaussées*. In 1715, he published his *Traité de la Construction des Chemins en France*, and in 1716, his greater work, *Traité des Ponts*, which, reprinted several times, remained a standard work on the subject for almost seventy years. The practical nature of the book was its main attraction although, by his caustic comments on theoreticians, Gautier accelerated the pace of those engaged on the advance of structural theory and possibly persuaded them to make their pronouncements more understandable for the practising engineer.

The great flowering of French technological science suffered a severe set-back during the French Revolution. Professors and students were suspect of counter-revolutionary activities, so the universities and schools were closed by the Government. France, however, in the throes of revolution, was also engaged in war and lacked military engineers; this brought into prominence Gaspard Monge, a great mathematician whose political views did not conflict with those of the Government. Monge recruited the necessary scientists and engineers to form, in 1795, a new school, the École Polytechnique. Here, for the first time, engineers were given a broad education in the sciences fundamental to their profession and from this school went students fully equipped for the advanced studies of the École des Ponts et Chaussées and their

Gaspard Monge *invented a new branch of mathematics—descriptive geometry—which was of such importance that the French kept it as a military secret for a quarter of a century.*

ultimate responsibilities. Gaspard Monge was born in Beaune in 1746, and from an early age showed great aptitude in mathematics, which enabled him to enter the military school at Mézières; his progress there was enhanced by his gift for teaching, and led to his appointment as professor of mathematics. While at Mézières, Monge invented a new branch of mathematics, called by him descriptive geometry, which was of such practical importance that for a quarter of a century the French kept it as a closely guarded military secret. It is now an essential part of the technical equipment of every engineer under the name of engineering drawing, the organized method of placing plans, elevations and sections on a two-dimensioned sheet to represent three-dimensional objects. This discovery was taught to young officers of the French services and in that way became known to a young naval officer, Marc Isambard Brunel, who later became a refugee from France during the Revolution. After a few years in America, Brunel came to England, where his knowledge of this new method of communicating engineering ideas found a fruitful field. There seems little doubt that Brunel's importation of the invention of Gaspard Monge enabled British engineers to exploit their profession in the most effective manner during the nineteenth-century expansion.

Monge became a great friend, possibly the only real friend, of Napoleon Bonaparte, who used the scientist's advice on all matters of education, commerce, science and engineering; in fact, Monge was to Napoleon in a much greater degree what Lord Cherwell (Professor Lindemann) was to Churchill a century and a half later. Following Napoleon's downfall, Monge fell from grace and died in the slums of Paris on 28 July 1818, expelled by the Academy, disgraced by the Bourbons, but rightly idolized and honoured to the end of his days by the grateful students of the Polytechnique.

The end of the eighteenth century may be said to represent the peak of French ascendancy in engineering science; although it was by no means dead, the output of scientific work from the French schools was to influence civil engineering during the whole of the nineteenth century. In Britain, however, the great industrial

expansion led to an increasing interest in the best means of carrying out the engineering works of substantial magnitude required for the new systems of communications, particularly the railways. For these, bridges of great span were needed for carrying heavy loads at speeds hitherto unthought of. This led to a revival of interest in truss construction, an art of ancient origin. From the evidence of the bas-reliefs of the Trajan Column in Rome, the engineers of that emperor used timber trusses in the bridge over the Danube. The subject revived again during the Renaissance, when the Italian architect Palladio designed bridges of 100 feet or more span in timber truss construction. The working carpenter reached the pinnacle of his craft with the building by Jean-Ulrich Grubenmann in 1757 of a timber truss bridge with spans of 171 feet and 193 feet over the Rhine at Schaffhausen with a safe working load of up to 25 tons. The same carpenter and his brother in 1778 built a bridge over the Limmat near Wettingen having a span of about 390 feet. Railway construction in the developing countries of America and Russia called for bridges of simple and inexpensive design using local materials, and truss construction in timber offered a solution. By 1840, a number of truss designs, based on the use of timber, had been evolved in America; in that year the railway engineers of that country trans-

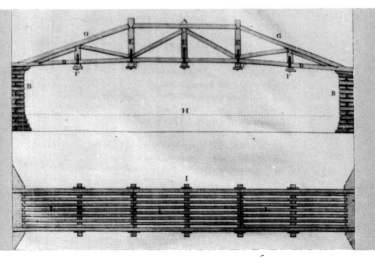

Timber truss construction *was used in Trajan's bridge over the Danube in* A.D. *104* (*left*) *and was revived during the Renaissance by the Italian architect, Palladio.*

Typical of truss design methods *in the middle of the 19th century was th*

ferred their attention to trusses in cast iron and wrought iron. S. Whipple, in his *Essay on Bridge Building*, published in Utica, New York, in 1847, laid the foundations of a new art of design based on the use of analytical and graphical methods of solving the problems of determinate trusses. His graphical method included the drawing of a force polygon for each joint of the truss.

Crumlin Viaduct on the Newport–Hereford railway.

In 1846, Warren introduced his well-known truss and by 1850, the methods of truss design were well known in Europe and America.

Before any phenomenon in the physical world can be studied scientifically it is first necessary to state it in terms which may be measured. This, for the elastic behaviour of materials, was done

331

Many advanced ideas *put forward by Thomas Young were not noticed in his time and credit for them was not given until much later in the 19th century.*

by Thomas Young, who was born in 1773 to a Quaker family living at Milverton in Somerset. He was a remarkably precocious child and by the time he was fourteen had a working knowledge of mathematics and of a number of languages sufficient to enable him to earn his livelihood as a tutor to the children of a rich family. While working in this capacity, he was able to continue his studies and, in 1792, he began to study medicine, in London, Edinburgh, and, finally, at Göttingen University, where he obtained his doctor's degree in 1796.

He returned to England in the following year and entered Emmanuel College, Cambridge, as a Fellow Commoner, studying sound and light and, in 1801, discovered the phenomenon of the interference of light. In the following year he was elected a member of the Royal Society and became professor of natural philosophy at the Royal Institution where he remained until 1803. In 1807, he published the material of his lectures and in them he included the idea of a modulus of elasticity for the first time, although not defined in the terms now accepted for what we know as Young's Modulus. Young's lectures, as published were difficult to understand; and for that reason many of the very advanced ideas put forward were not noticed in his time and credit for them was not given to him until much later in the nineteenth century.

The railway age brought other great teachers and co-ordinators of the science of engineering structures. One of the greatest of these was W. J. McQuorn Rankine, born in Edinburgh in 1820, the son of an army officer who, on retirement, had adopted railway engineering as a career. Rankine was a graduate of Edinburgh University who gained practical experience on railway construction in Scotland and Ireland. In 1843, he presented a paper to the Institution of Civil Engineers on fatigue failures of railway axles in which he laid great importance on the need for a radius joining any change of section. Rankine became a prolific author of papers on scientific subjects and was elected F.R.S. in 1853; this was followed by his appointment in 1855 to the chair of engineering at Glasgow University, the first university to adopt engineering as a main subject. He remained at Glasgow for the rest of his days, devoting his life to the education of engineers, not only in his own faculty, but, through his writings, in all parts of the world. In his two books *Manual of Applied Mechanics* published first in 1858 and *Manual of Civil Engineering* published in 1861, Rankine co-ordinated the loose pieces of scientific knowledge, tidied up the subjects and added much that was his own; they have passed through many editions and may be read with advantage even today. The drawback of Rankine's writings is his style, which is difficult to read, a vice which has been followed by many writers of textbooks down to the present day. His principal contributions to the theory of structures were made in respect of the application of statics to structural design, particularly of arches; he laid the foundations of the application of reciprocal diagrams which led to the work of Clerk Maxwell on reciprocal figures, later made a practical working procedure by Robert H. Bow. Rankine's investigation into the stability of loose earth was, however, his greatest contribution to civil engineering, providing as it did the start of a study into the behaviour of soils leading to the present rapidly growing science of soil mechanics.

The need for bridges of increasing span led to revival of interest in suspension bridges which, although of ancient origin, had received little scientific investigation before the work of Navier. This great engineer, whose principal work was on materials, was born in Dijon in 1785, the son of a lawyer. C. L. M. Navier lost his father at fourteen years of age and became a ward of his uncle, E. M. Gauthey, a famous engineer and student of Coulomb, who educated the boy as an engineer via the École Polytechnique and the École des Ponts et Chaussées. Gauthey died in 1807, a year before his nephew graduated, leaving unfinished a treatise on bridges and channels. Navier undertook the completion of this work; between 1809 and 1816 he published it in three volumes with editorial notes of his own. These notes are of great interest to the historian as from them can be assessed the state of knowledge in engineering theory, especially that very considerable part due to his own work. The French Government, interested in the work of British engineers, especially that of Thomas Telford, who was building the Menai Bridge, sent Navier to Britain to study these developments and following two visits in 1821 and 1823, he presented in 1823 his *Rapport et Mémoire sur les Ponts Suspendus*, a masterpiece of historical, descriptive and theoretical writing on the subject. It remained the standard work for half a century and is of importance even today.

In 1826, Navier published his book *Leçons sur l'Application de la Méchanique*, a general review of structural theory and the strength of materials. His contributions to structural theory given in this book include methods for determining the deflection curves of beams and cantilevers under uniform or point loading; a general method of analysing statically indeterminate problems in mech-

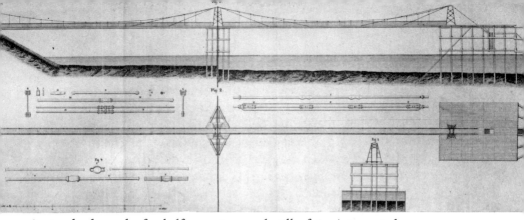

A standard work *for half a century and still of importance today,*
Navier's 'Rapport et Mémoire sur les Ponts Suspendus' contains this
design for a bridge.

anics of materials, such as those presented by built-in beams; the
theory of bending of curved bars; and new work on thin shells,
arches, plates and trusses. Navier's work on continuous girders
was too complex for day-to-day use by practical designers, depending as it did on the solution of each span separately, followed
by a comparison of the solutions in order to determine the con-
stants of integration. His methods were made applicable to many
more problems by the discovery of B. P. E. Clapeyron, who
observed that the bending moments at three consecutive supports
are connected by an invariable relationship. Clapeyron used
expressions for the angles made by the tangents to the deflection
curve at the supports with the initially straight axis of the beam,
and from them obtained a number of equations equal to the
number of unknown quantities, all of which can be found.
Although Clapeyron used his method in 1849, or earlier, he did
not publish it until 1857. The modern form of Clapeyron's
equations, the 'Theorem of Three Moments', was first published
by Bertot, who made the original theory more straightforward
to work out, but the ultimate practicability of solving the problem was made feasible by Otto Mohr, who evolved graphical
methods which for some years superseded the elaborate calculations.

335

Otto Mohr was born in 1835 at Wesselburen in Holstein on the North Sea coast. He became a structural engineer on railway work and as a result of his reputation as a designer of bridges, in 1868 became professor of engineering mechanics at the Stuttgart Polytechnic where he remained until 1873. He then went to the Dresden Polytechnic and continued teaching until his retirement in 1900. His last years until 1918 were spent on his scientific work. Mohr introduced the principle of virtual displacements into the solution of framework problems; he greatly simplified the methods of solving problems of beam deflections and was responsible for the introduction of influence lines into the design of engineering structures. His work on the strength of materials led to the development of graphical methods for determining the stress at a point; this led to a great advance in knowledge of the subject as worked out by him and others. The well-known Mohr circles are used not only in the solution of problems of structures in normal engineering materials, but also in the important and rapidly developing field of soil mechanics, to which Mohr's work has important applications.

Mohr was greatly influenced in his ideas by the work of James Clerk Maxwell, born in Edinburgh in 1831, who in his early youth, showed great aptitude in mathematics. At fifteen years of age he devised the method of drawing an ellipse with the use of two pins and a thread and presented a paper on the subject to the Royal Society of Edinburgh. Maxwell's greatest contribution to structural theory was his work on the elastic properties of frameworks, leading to the reciprocal relationships for deflections, applicable to all linear elastic structures, and enabling him to solve statically indeterminate structures by influence coefficients. His paper in the *Philosophical Magazine*, 1864, was later developed by Muller-Breslau in 1886. Another of Maxwell's contributions to engineering structures began in 1847, when he started his investigations into the properties of polarized light; from these studies he developed his methods of photoelastic stress analysis. In 1850, he presented a paper on the subject to the Royal Society of Edinburgh, in which he described his photoelastic method applied to a number of problems of stress analysis, using moulded

James Clerk Maxwell *had a wide range of scientific interests, including hydrostatics, optics, astronomy, electricity and magnetism, in addition to his work on elasticity and the theory of structures.*

isinglass jelly as his material. The originator of this method of stress investigation was D. Brewster, who published a paper on it in 1816. It is based on the experimental fact that an isotropic transparent body, when stressed, becomes doubly refracting, with its optical principal axis at any point in the directions of the principal axes of stress at that point. James Clerk Maxwell had a wide range of scientific interests including hydrostatics, optics, astronomy, electricity and magnetism, in addition to his work on elasticity and the theory of structures. His most famous scientific paper was probably that written while he held a chair at King's College, London, between 1860 to 1865, in which he enunciated the electromagnetic equations from which the whole practice of radio-communication has developed. His last years were spent in directing the great new physics laboratory at Cambridge, named Cavendish after the donor of its cost, the Duke of Devonshire. In that capacity he greatly influenced the policy of the university with beneficial effects to science and engineering which extend to the present day. He died in 1879. Cambridge is also closely connected with the work of Lord Rayleigh (John William Strutt)

337

who, in the second half of the century, added substantially to the knowledge of problems on vibration and of the stability of elastic systems. Much of his work was published in his two-volume *Theory of Sound* in 1877, and whilst much of this was concerned with fundamental acoustics, his studies of vibrating structures led to the development of techniques cf approximate analysis based upon concepts of minimum energy, using Euler's calculus, which have continued to be developed in later times, first by Ritz, and later by members of the Russian school of engineers, such as Galerkin and Kantorovitch.

Just as Navier, in his *Leçons sur l'Application de la Méchanique*, had pulled together the loose pieces of structural knowledge of his time, so, in due course, another scientific writer and investigator brought Navier's work up to date, adding, at the same time, great contributions of his own. This was Barré de Saint-Venant, who was born in 1797 in the province of Seine-et-Marne, the son of an agricultural economist. At sixteen, he entered the École Polytechnique, showing great promise in his early examinations. In 1814, however, Paris was threatened by the allied armies and the students, including Saint-Venant, were mobilized. While moving the guns into the fortifications, Saint-Venant, overcome by conscientious scruples stepped from the ranks and refused further duty for the usurper. For this, he was proclaimed a deserter and expelled from the Polytechnique. From then, he worked in the powder industry until 1823, when he was permitted to enter the École des Ponts et Chaussées without examination; even there, he was ostracized for the first two years, but graduated from the school at the top of his class. It is interesting to speculate whether such an unauspicious start to the career of a genius has, in the long run, any real effect on his life work; certainly, from this time nothing could stop Saint-Venant. He worked for some years as an engineer on river and canal improvements, while at the same time doing theoretical work. In 1834, he presented two papers to the Academy of Sciences; these created a great impression, and, as a result, he was invited in 1837 to lecture on the strength of materials at the École des Ponts et Chaussées. In these lectures he refers to many of the problems which he was, in his later work, to solve. Navier had been one of Saint-Venant's teachers and his book on mechanics was still the most advanced available, although somewhat out of date. Saint-Venant, in revising it, added so much new material that the new edition represented 90 per cent of Saint-Venant's notes to 10 per cent of Navier. This volume was published in 1864, but in 1853 Saint-Venant presented his most important paper to the French Academy; nominally on torsion, the paper reviews and co-ordinates all the theory of elasticity known at the time. His rigorous solutions to problems of torsion and bending, applied to practical examples, put a new impetus to the application of elastic theory to engineering structures, and from the publication of his revision of Navier's book, engineering textbooks took on a new look. He died on 6 January 1886, four days after the publication of his last article, and greatly respected by the engineers and scientists of the world.

The stability of retaining walls and foundations had always been of great concern to civil engineers, and, from time to time, attempts were made to advance the knowledge of the subject. Coulomb and others had developed the theory, but it was left to Alexandre Collin (1808–90) to make the first serious field investigations into the behaviour of soils, under conditions of instability. He made a thorough survey of some fifteen slips in the clay slopes of railway cuttings, embankments and earth dams and recorded his results in a treatise entitled *Experimental researches on Spontaneous Landslips in Clay Soils* which was published in 1846. In this book he describes, not only the approximately cycloidal form of the slips, but also a shear box which he had constructed for tests on clay samples. He also arrived at an approximate method of analysis of stability which was an anticipation of the $\theta=O$ analysis. In spite of its importance to engineers engaged in solving earthwork problems, the book remained almost unknown for about seventy years, indeed until after the subject of soil mechanics was vigorously restored to organized study by Terzaghi in the early years of the present century.

Towards the end of the nineteenth century, the methods of Clerk Maxwell had been adopted as a normal procedure for the determination of stresses in triangulated frames, largely due to the simplified means outlined by Robert H. Bow in his book *Economics of Construction in relation to Framed Structures*, published in 1873. The simple methods clearly explained by Bow applied only to fully triangulated, or perfect, frames; any under- or over-braced frames were regarded at that time with great distrust; although the friction of the riveted or bolted joints introduced a substantial degree of indeterminacy, which was, however, ignored. In Britain, particularly, Maxwell's work on the solution of problems in indeterminate structures was ignored, even by writers of textbooks until 1895, when *Engineering* printed a series of articles by H. M. Martin, afterwards published as a book, based on the work of a young Italian engineer, Alberto Castigliano. Born in Aste in 1847, Castigliano spent his early years as a teacher and, in 1870, entered the Turin Polytechnic Institute, where he did outstanding work on structural theory. He presented a thesis in 1873 for his degree in engineering and in that, stated the theorem which immortalizes his name This was presented two years later in an extended form to the Turin Academy of Sciences. In this thesis Castigliano showed that the strain energy of any component in a loaded system may be shown in terms of a horizontal and a vertical displacement and an angular rotation. By applying this theorem to the analysis of trusses, he proved the principle of least work. Castigliano died in 1884, but his ideas have been developed by succeeding engineers; notable among these were H. Muller-Breslau and E. Betti, who generalized the relationship between external work and strain energy. H. Manderla analysed the secondary stresses in a truss with stiff joints by the slope-deflexion method, that is, by treating the angles of rotation of the joints as the unknowns to be solved, rather than the moments and the direct forces; this innovation was published in 1879 following some years in which Otto Mohr and James Clerk Maxwell had also made valuable contributions to knowledge of the subject. Following the work of Maxwell and Muller-Breslau, Ostenfeld, a Danish engineer, published in 1924 a study of the elastic theory of structures demonstrating, for the first time, the fundamental nature of, and the duality relationships between, the force and displacement methods of structural analysis. Sur-

The use of models *as aids to structural design provides the means of assessing the correctness of shape and dimension in structures which may otherwise be difficult to analyse. A model of a section of the Medway Bridge is shown under test in the laboratory.*

prisingly, little notice was taken of this work, certainly in Britain and America, until after the Second World War, when the advent of the electronic computer forced a fundamental approach upon engineering analysis, and a revival of interest in Ostenfeld's work. Maxwell's work on photoelasticity also had a direct bearing on the strain-energy approach which led to the more recent researches of E. G. Coker and Filon and to the present-day extensive use of photoelastic methods of stress analysis. The slope-deflexion method of analysis is amenable to confirmation by models, a method investigated by Professor G. E. Beggs, whose paper on the subject was published by the Franklin Institute in 1927. Beggs' method is valuable for problems which are too complex for exact mathematical treatment, but the advent of the electronic computer has to some extent provided a means of approach to many problems of mathematical complexity, particularly those involving laborious arithmetical operations. The use of models as aids to structural design has, however, developed very considerably and, with such equipment as multi-channel strain gauge recorders, provides the means of assessing the correctness of shape and dimension in structures which may otherwise be difficult or even impossible to analyse.

Methods of successive approximation have been developed for the solution of involved frame and similar structural problems. Of these, probably the most important is the moment-distribution method of Professor Hardy Cross, introduced in his paper to the American Society of Civil Engineers in 1930. The moment-distribution method starts with the assumption of complete fixity at every joint and proceeds alternately to balance and distribute the moments until the required degree of accuracy is achieved. The generalization of methods of successive approximation is due to Sir Richard Southwell who, in 1940, published his *Relaxation Methods in Engineering*.

Just as in earlier periods of rapid development, teachers have come forward to consolidate the gains by their writings, so in the present century a few gifted men have brought order into the subject of structural theory. Of these, the names of J. C. Maxwell, A. E. H. Love, S. P. Timoshenko and K. Terzaghi are outstanding. The first three, by their personal contributions to new knowledge in the behaviour of materials in structures and the fourth for his pioneer work in soil mechanics. All four, by their writings, have co-ordinated the work of others and provided sure foundations for later work. Like their predecessors of earlier centuries, they provide sufficient proof that research and teaching may be, and must be, combined. Thus is provided the inspiration of each new generation which gains its own basic learning from those working on the frontiers of new knowledge.

Sub-division of labour — basic techniques developed by the Greeks and Romans — Vitruvius — pumps, cranes and coffer-dams — the advent of steam power Brunel's Thames tunnel — compressed air and the diving bell — steam pumps and excavators the internal combustion engine — the tracked vehicle

◄Forty horse-driven capstans, *with multiple lifting blocks, were used by Domenico Fontana in 1590 to re-erect the obelisk, brought from Heliopolis in* A.D. *37, in St Peter's Square at the command of Pope Sixtus V.*

Construction Methods

MAN'S INCREASING KNOWLEDGE of his environment, and his desire to alter it to suit his comfort and convenience, has led, from the earliest times, to the invention of new methods of construction. These were based, for many thousands of years, on the use of his own muscle power. The earliest tools, whether for excavating earth, hewing wood or moving weights were based on the energy of man, either alone or in groups. The earliest evidences of civilization are to be found in the remains of man's early civil engineering works. In these, the magnitude of the effort indicates that large numbers of people were employed, often for long periods, implying that, for the continued existence of these workers, others must have been employed on the production, storage, and distribution of the food and other necessities of bare living. All this needed leadership, and from these essentials, an organized society tended to develop. In such a society, the specialist would appear, whose work, based on special skill, would justify his full time employment at one kind of work; thus the carpenter, mason and worker in metals would be of more value to the community than the unskilled labourer and would demand privileged treatment, while the gifted persons who organized the whole would demand even greater privileges, so that some stratification of society must have developed.

By the time the early civilizations of the Middle East had matured, the sub-division of labour had so developed that works of great magnitude could be undertaken, using for the heavy work, vast numbers of labourers, either conscripted from the peasantry or captured in battle. The cutting of great watercourses and the raising of the vast religious monuments by the ancient Egyptians must have involved the expenditure of many thousands of lives; early writers estimate that 120,000 died in an attempt to anticipate the Suez Canal in an excavated cut which reached the Bitter Lakes. The major contribution of the Egyptians to civil engineering arose from the annual inundation of the fertile lands of the Nile valley, which removed the evidences of boundaries between properties. This led to the need for surveyors who were able by their art not only to re-establish the boundaries accurately, but through their acknowledged competence to satisfy the land-owners that this was so done.

With the Greek civilization came an economy, based on slave labour, which encouraged the growth of a professional class of people who, having leisure to think and discuss their ideas with others, greatly advanced the principles of construction, and, in some respects, provided the basis of our modern construction methods. The titles given to this class of practitioner include that of *Architecton*, or chief technician and *mechanicos*, or mechanician. The works of these people are described by writers such as Herodotus who, in his *History*, described the construction of a tunnel in Samos, formed by driving headings from both sides of a mountain and requiring over 3,000 feet of rock cutting. Although the headings did not meet exactly, the error of line was only about 20 feet and in level only half that amount. Such an error indicated surveying methods of some degree of sophistication. The Greek mechanicians Hero and Archimedes, whose writings so greatly influenced later engineers, were clearly themselves influenced by their own predecessors and contemporaries— such men as Chersiphron who, by encasing the columns for the Temple of Artemis in wooden drums, was able to roll them in one piece a distance of 8 miles from the quarry to the site. The lifting of pieces of masonry weighing 50 tons or more to form girders or lintels at a height of 50 feet was achieved by methods based on the principles of the lever and the inclined plane.

The Romans, while not adding appreciably to theoretical principles, were, as practical engineers, able to exploit fully the knowledge of mechanical principles which they had learned from the Greeks. Benefiting from military success, their economy also possessed great resources of manpower and materials, which were

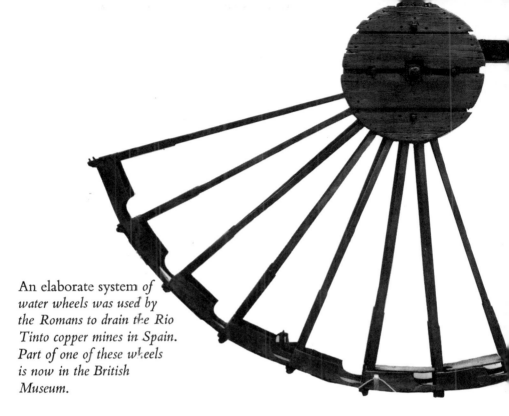

An elaborate system *of*
water wheels was used by
the Romans to drain the Rio
Tinto copper mines in Spain.
Part of one of these wheels
is now in the British
Museum.

lavished on their massive works of engineering. By occupying
the countries which formed their vast empire, they were able to
draw on resources of skill and knowledge which, under their own
highly organized society, resulted in a material advancement in
the arts of construction. Although manpower was, at first, the
basic unit on which all operations depended, the individual effort
was multiplied by the use of mechanical aids, such as treadmills
on which up to fifty or so men could apply their effort, by multiple lifting blocks, applications of the lever and the inclined
plane. By the time the Roman influence had reached its height,
about 70 B.C., the use of water as a power supply through the
water wheel had been developed and many applications on fixed
sites were effectively powered with falling water. Many Roman
writers refer to technical aspects of Roman life, but the greatest
contribution to our present-day knowledge of Roman engineer-

347

ing is due to the writings of Marcus Vitruvius Pollo who flourished about 20 years B.C. In his book *De Architectura*, Vitruvius describes the general state of the mechnical and constructive arts of his day and his work is supplemented by that of Sextus Julius Frontinus who, a century later, wrote a full account of the water supplies of Rome. The written descriptions of these two authors is supplemented by the bas-relief sculptures-on the column erected in Rome to commemorate the emperor Trajan. The monument, over 125 feet high, has on its shaft a sculptured representation in the form of a spiral, of the activities of the emperor during his lifetime, and as these include many of his achievements in engineering, the details are of great value as a pictorial record.

From these sources, and from a study of their works as we find them today, we learn that the Romans organized engineering works as military or civil requirements dictated. In the military stages of occupation of a country, their engineers would rely on the services of the common soldiers and of captured enemies for manual labour. This stage would continue until the country was fully settled, when the civil government would administer the necessary works of construction, many of which would be carried out by contractors under terms and conditions very similar to those of our own time. Roman methods of surveying and setting out work were essentially simple; these were based on the use of measuring rods and the *groma*, an instrument for setting out lines at right angles. Much of their use of straight lines and right angles for planning would appear to be due to the inability of most of their surveyors to work in any other layout. Vitruvius makes it clear that Roman methods of lifting heavy weights by blocks and tackle, or by levers, were highly efficient. It is also evident that many of their mechanical contrivances were derived from the Greeks, especially in the matter of pumps; the screw of Archimedes and the plunger pump of Ctesibus were used for many applications. In their harbour and bridge works, the Romans used coffer-dams, sometimes founded to a considerable depth, and their pile driving equipment, although dependent on manpower, must have been fully effective for its purpose.

Roman bridge foundations on rock were constructed in coffer-dams; these were built with double skins of stakes, tied and propped for stability and filled with clay packed in wicker baskets, the baskets substantially reinforcing the clay fill. The water was pumped out, usually with Archimedean screws, during the period of excavation and construction.

For many centuries after the Romans, construction works were dependent on human effort and little advance was made. In a few cases, men were replaced or assisted by animals, but the basic methods remained substantially the same until the advent of two new factors, steam power and the use of explosives. Both of these, by reducing very substantially the number of men employed, made works of greater magnitude economically possible and, by shortening the time of construction, encouraged promoters of schemes to undertake works of which they were able to forsee completion within their own lifetime. The improvements in materials which made the steam engine possible, also enabled the civil engineer to invent new physical and mechanical aids to expedite his work. The methods used by the French engineers of the eighteenth century, although more refined, differed little from those of the Romans; the new developments followed the techniques of the iron-using industries which, advancing rapidly in Britain, in the nineteenth century carried civil engineering progress to that country. The ancient crafts of the carpenter, mason and smith, were now to be supplemented by those of the ironfounder and engine builder.

One of the first to take advantage of the new iron technology was an *émigré* Frenchman, Marc Isambard Brunel, who with his tunnelling shields, built of cast-iron sections, made it possible to tunnel under the Thames in materials hitherto impossible. Like all pioneers, Brunel had to contend with long periods of frustration owing to difficulties arising from lack of experience, prejudice and financial stringency; this delayed the work which, from the formation of the original company in 1823, took until 1842 to complete. Brunel's tunnel was preceded by an attempt by Richard Trevithick to drive a drift under the river from Rotherhithe. This he succeeded in doing for a distance of 1,046 feet, the section

Little or no advance *was made in methods of construction from the fall of the Roman Empire until the invention of steam-driven machinery which made obsolete such man-powered machines as these. Perronet's 18th-century crane (right) hardly differs from the 16th-century example below.*

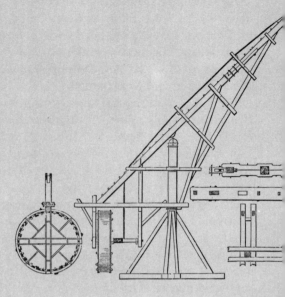

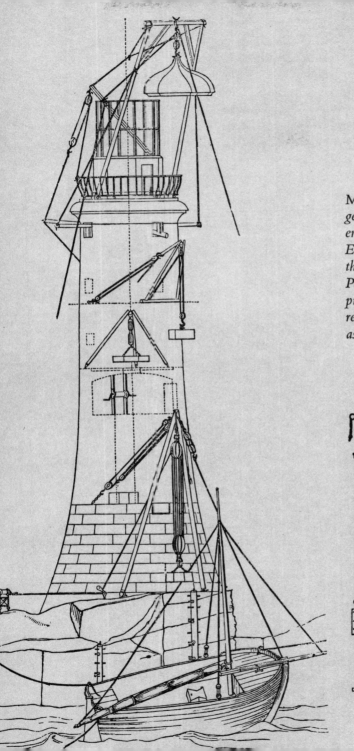

Man-powered *lifting gear was used in the erection of Smeaton's Eddystone lighthouse in the 18th century. Perronet's method of pile-driving (below) required the services of as many as twelve men.*

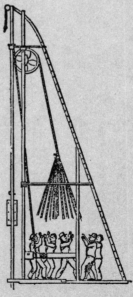

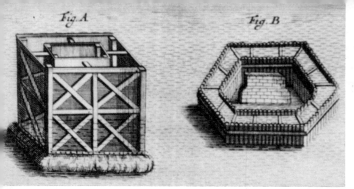

Coffer-dams *were employed by the Romans in the construction of their harbours and bridges. Above are two examples from the 18th century, below a coffer-dam built to repair Blackfriars Bridge in 1836.*

An authentic record, *made by the resident engineer, of the building of Rennie's London Bridge in the 19th century, showing the coffer-dam, elaborate wooden centering as well as a variety of lifting gear. Temporary works such as these are as much a part of the engineer's art as the finished product itself.*

A modern coffer-dam *built for the construction of a tunnel 'in the dry' under the North Sea canal at Velsen, near Amsterdam.*

354

being 5 feet high, 2 feet 6 inches wide at the top and 3 feet at the bottom. Early in 1808, the workings were flooded out by a break-in of the river and the work was abandoned. Brunel excavated his double tunnels through the London clay from Rotherhithe to Wapping, starting from a 50-foot diameter shaft on the Rotherhithe side sunk to 65 feet depth. The shield consisted of twelve cast-iron frames, each of three cells in which the excavation was done under the protection of poling boards advanced into the clay by screws while each of the twelve shield sections could be advanced in turn by jacking against the brick tunnel

lining. Although the Thames broke through to the workings on three occasions, on one of which Brunel's son, Isambard Kingdom, nearly lost his life, the engineer pursued his object with resolution and, in spite of the works being completely suspended from 1828 to 1835, he saw its successful conclusion seven years later. The Thames tunnel alone was sufficient to give Marc Isambard Brunel a high place in the history of engineering, but, when combined with his other works, it must surely put him equal to the greatest engineers of all time and even greater than Isambard Kingdom Brunel, his brilliant and more famous son.

Brunel lacked one essential for successful tunnelling under water—the use of compressed air. The diving bell, in the form of a box, open on the underside, suspended in the water from a crane, had been in use from an early date for men to work under water and had been progressively improved. When the apparatus suitably ballasted, was sunk to a depth, the air contained in it was compressed by the pressure of the surrounding water and its volume reduced. If, therefore, air under pressure could be added, the duration of work could be extended according to the amount of air introduced. In 1721, Halley, the astronomer, proposed to supply air in bottles which, being released in the bell, would fulfil the purpose. Coulomb proposed, and Smeaton achieved the same purpose with air pumps and by the end of the eighteenth century, the diving bell was an accepted method for performing work

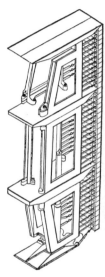

The historic Thames tunnel *built by Brunel is now used by the railway. The shield (right) made tunnel driving much easier. The platforms were moved forward within the protection of the shield as the earth was removed.*

under water. Its disadvantage was that when the bell was submerged, it was not possible to pass men or materials in or out of it without them passing first through the water. In 1830, Admiral Thomas Cochrane, Earl of Dundonald, took out a patent for the use of compressed air, which incorporated a fundamental principle of great importance—the air lock. In Cochrane's invention, which he envisaged as specially applicable to tunnelling, the working space was sealed off by a diaphragm which was provided with a small entrance chamber having two doors both opening inwards, one from the atmosphere and one to the pressure zone. Men and materials entering the chamber, or air lock as it became, closed the door from the outside and turned on air to balance that in the working space; when the two pressures equalized, the inner door, previously held by pressure, could be

357

opened and the men could pass into the chamber, the outer door being held tight until the process was reversed. Cochrane's invention opened up great possibilities in underwater work and its first applications were successful; a mine shaft was sunk in France by Triger, a mining engineer, with its use in 1839, and, in 1851, I. K. Brunel, using compressed air, sank the central pier of his bridge at Chepstow carrying the South Wales Railway across the Wye. The most effective use of compressed air came with the use of a more advanced form of tunnelling shield designed by James Henry Greathead who, in 1869, first used it to cut the Tower Subway in London clay under the Thames. Greathead's shield combined with the use of compressed air was subsequently used in the construction of the tunnels for London's underground railway system and many other works. The availability of a convenient form of power supply in the form of compressed air was subsequently exploited by engineers to drive mechanical aids for such purposes as riveting, drilling, rock cutting, hauling and lifting. So successful did this medium of power transmission become, that in most civil engineering works undertaken today, some economic application of compressed air is to be found, while in mining the air compressor is now one of the most important items of equipment.

The use of steam power for civil engineering purposes was an almost inevitable outcome of the construction of the railways; without it, such works as Robert Stephenson's Kilsby Tunnel might have been impossible. At Kilsby, the use of steam for pumping enabled the work to continue to a final conclusion and, as at Kilsby, most of the early applications of steam were for pumping. Steam power applied to winches proved more economical than the best means previously available, such devices as the horse gin, treadmill and capstan, which had enabled larger multiples of power to be applied to a shaft, while multiple blocks had

Using compressed air, I. K. Brunel sank the central pier of his railway bridge at Chepstow in 1851. In this photograph taken in 1962, the original bridge is being replaced on Brunel's piers by a deck bridge of Warren truss design.

360   The construction of Kilsby Tunnel, *on the London–Birmingham Railway,* *caused endless trouble. In this lithograph by John Bourne, steam pumps are at*

work removing flood water from the tunnel workings, while horse gins raise excavated material to the surface.

performed the same function in lifting. Forty horse-driven capstans with multiple lifting blocks were employed by Domenico Fontana in 1590 to lift the 327-ton obelisk on the Piazza of St Peter in Rome; capstans were still used when Telford's Menai Bridge chains were hauled into place. Horse gins had been extensively used in conjunction with inclined planes to assist excavation on such works as the great railway cuttings. The steam engine enabled such large forces of men to be dispensed with at the cost of an incredibly small consumption of coal. Before long, steam was being used more directly as in James Nasmyth's steam pile driver of the 1840s and, under the pressure of manpower shortage during the great railway expansion in America, the steam excavating machine came as a natural development.

The steam excavator greatly expedited the great canal works which were coming into being towards the end of the century and, indeed, made such works economically possible. Manpower was becoming more expensive, labour had organized itself and no longer was it possible to hire vast armies of pick and shovel workers. In addition, time had become a vital factor in affairs, the great capital sums required for major works could not be tied up for ten or twenty years without interest; shareholders demanded speedy results and the machine could provide the answer. One of the first works on which machine power was used on the grand scale was the Manchester Ship Canal. The material removed to cut this waterway was estimated to be $53\frac{1}{2}$ million cubic yards, of which 12 millions were in sandstone rock. The man responsible for this great muck-shifting operation was A. O. Schenk who, in addition to a fleet of dredging plant, introduced no less than ninety-seven steam excavators to remove the greater part of the material 'in the dry'. The steam navvies, as they became called, removed 50 million yards of material, all of which was taken away to the spoil grounds by railway in 6,300 trucks drawn by 173 locomotives.

The principal excavating tool was the Ruston and Dunbar Steam Navvy, of which fifty-eight were employed. This machine consisted of a rectangular wrought-iron frame carried on rail wheels, supporting an engine and boiler at one end. The crane

The primitive equipment *used by the first railway engineers contrasts strangely with their massive undertakings. This scene is at Edgehill Tunnel on the Liverpool–Manchester Railway.*

jib, built in two parts, was pivoted on a wrought-iron tower; between the two halves of the jib was an adjustable arm carrying the scoop or bucket. Two men were required to operate the machine, which had a capacity of up to 1,000 yards per day, 600 yards being a fair average for a ten-hour day. This heavy piece of equipment was self-propelled on its railway track, but its cut was limited by the radius of the jib, so that it was usually operating to cut a trench for itself in a forward direction. For a wide cut, such as the Canal, a number of such excavators worked in echelon, each with its own series of tracks for itself and its attendant spoil wagons.

The cleaning up of odd places was performed by lighter steam excavators which were developments of the ordinary steam crane of the time. These had a 360-degree turning circle and could therefore, if necessary, cut and load in any direction. For the side slopes, excavators of French and German origin were used. These were similar in design to the floating bucket ladder dredger and were able to cut, load and travel continuously, the output averaging 1,500 cubic yards per day in soft material.

Steam plant also eased the problem of the harbour engineer in the building of breakwaters with large stones. The earliest works of this kind were executed by engineers who had no alternative but to tip the heavy pieces at random from the deck of a barge or to lower it from caissons. The mole at Tangier, constructed during the reign of Charles II was built in this way. In June 1677, Henry Shere, the engineer in charge, floated out and deposited on a rubble mound, the first of a series of great monolithic blocks weighing over 2,000 tons. This block, duly christened 'King Charles', was 42 feet long, 42 feet wide, and 18 feet high. A few days later a 600-ton block was placed in position and three more were almost ready to deposit. These blocks were laid on the rubble mound 6 feet below low water and extended 3 feet above high water, where a parapet, 10 feet thick and 10 feet high, made the total height of the breakwater about 60 feet above the sea bed.

Machine power *on the grand scale was used in the construction of the Manchester Ship Canal. The principal excavator was the Ruston and Dunbar Steam Navvy (right), nick-named 'Jumbo', of which 58 were employed.*

The contract was originally let by Charles II to the Earl of Teviot, Sir John Lawson, and Sir Hugh Cholmley, on 20 November 1662, at 13 shillings per cubic yard; three years later the price was increased to 17 shillings and as that proved insufficient, the contract was cancelled and put into the hands of the contractor's agent, Henry Shere, who undoubtedly was one of the great engineers of his time.

Unfortunately, this great work came to an untimely end as, twenty-one years later, the king found that everyone but himself had been making money out of the Tangier North African station, the annual cost being about £70,000, mostly from his own pocket. He therefore decided to send a fleet, under Lord Dartmouth, to demolish the city, its forts, walls, and the massive mole. The inhabitants were to be compensated and to supervise this part of the operation, the king sent Samuel Pepys, his trusted and industrious servant. With Pepys went Will Hewer, his friend, and Henry Shere, the engineer. There was also a young ensign, Thomas Philips, who, in the later years became Second Engineer of England and one of the greatest military engineers of his time.

John Rennie used similar methods when building Plymouth breakwater, but his stones were limited to a maximum size of about 10 tons. These were carried to the site in special barges equipped with railway tracks on which the stones were carried in trucks, designed to expedite tipping from the stern of the vessel. During the period from 1812 and 1841, while the breakwater was under construction, steam entered the maritime world and the later stages of building were expedited by the use of steam tugs.

By the middle of the nineteenth century, large cranes were coming into use for handling heavy blocks in harbour construction. In 1869, a 'Goliath' type of crane of timber and iron construction was used in the block-yard at Karachi. This machine lifted a maximum load of 40 tons by a hydraulic cylinder which received its pressure from steam-driven hydraulic pumps, the same engines being available to move the crane in longitudinal and transverse directions. For block setting, machines variously called 'Titans', 'Mammoths' or 'Hercules' were developed by the specialist crane-building firms. These were able to set blocks up to 50 tons in weight to radii up to 100 feet and had motions for

Mechanical excavation *and lifting. Below is the 'Titan' crane used at Peterhead and on the right the 'Rapier 1200' excavator removing overburden from an opencast iron ore working.*

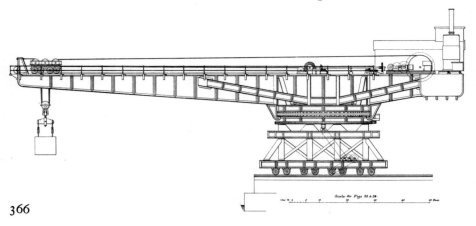

travelling, horizontal setting and slewing, as well as lifting. Such a crane was designed in 1890 for work at the Admiralty Works at Peterhead under the supervision of Mr (later Sir William) Matthews and constructed by Messrs Stothert and Pitt at Bath.

By the end of the nineteenth century, the use of concrete for mass work was made possible by the development of the power-driven concrete mixer. From the earliest days of their use, these machines were constructed in two principal types, the continuous mixer, and the intermittent, or batch mixer. In all the machines developed, the main requirement was that the material should be turned over sufficiently to produce a uniform product. Such mixers were available before the end of the century and the present great variety of concrete-mixing machinery represents the result of three-quarters of a century of refinement in detail, plus the advantage of modern portable prime movers, and not any radical change in the principles on which the machines operate.

The turn of the century brought two major innovations to civil engineering plant, the internal combustion engine and the tracked vehicle. The first of these provides a major chapter in the history of mechanical engineering, second only in importance to the coming of steam. Not only has it revolutionized communications, but in its wide range of powers from several thousand down to one horsepower or even less, it has provided the means of bringing power to almost any kind of civil engineering process. The tracked vehicle, developed at first for military purposes, provided the means of locomotion over soft and unreliable ground, a difficulty often met in civil engineering. Its application to civil engineering plant produced pieces of equipment such as the bulldozer, which now supplements modern, internal combustion-engined versions of the old steam navvy in speeding the age-old operation of muck shifting.

Selected Bibliography

The books included in the following list are selected from a great number referred to in the preparation of this history. Most of them are recent publications and likely to be in print while the remainder are probably to be found in most county or borough libraries.

GENERAL

A History of Civil Engineering, Straub, London 1952.
The Story of Engineering, Finch, New York 1960.
A Social History of Engineering, Armytage, London 1961.
A Short History of Technology, Derry and Williams, Oxford 1960.

BIOGRAPHY

Isambard Kingdom Brunel, Rolt, London 1957.
Thomas Telford, Rolt, London 1958.
George and Robert Stephenson, Rolt, London 1960.
Lives of the Engineers, 4 vols., Smiles, London 1862.

ROADS

The Road Goes On, Scott-Giles, London 1946.
The Story of the King's Highway, S. and B. Webb, London 1909.

RIVERS AND CANALS

Inland Waterways of England, Rolt, London 1950.
British Canals, Hadfield, London 1950.
Our Waterways, Forbes and Ashford, London 1906.

RAILWAYS

An Economic History of Transport, Savage, London 1959.
British Railway History, 2 vols., Ellis, London 1954–9.
English Railways, Cleveland-Stevens, London 1915.
A History of British Railways to 1830, Dendy Marshall, London 1938.

DOCKS AND HARBOURS

Ports and Harbours, Morgan, London 1952.
The Red Rocks of Eddystone, Madjalaney, London 1960.

WATER SUPPLY AND PUBLIC HEALTH

The Story of Water Supply, Robins, London 1946.
The Water Supply of Greater London, Dickinson, London 1954.

BRIDGES

Bridges of Britain, De Maré, London 1954.
The Story of the Bridge, Robins, Birmingham 1948.

MASTERY OF MATERIALS

Industrial Biography, Smiles, London 1863.
Autobiography, Bessemer, London 1905.

STRUCTURAL THEORY

History of the Strength of Materials, Timoshenko, New York 1953.
Men of Mathematics, Bell, Penguin Books.

Index

Acknowledgements

The author and publishers gratefully acknowledge the courtesy of the following who have granted permission to reproduce pictures:

Aero Films and Aero Pictorial Ltd, pp. 25, 256; Aerofilms Ltd, pp. 15, 18; Amsterdam Ballast Co., pp. 354–5; Anderson, p. 187; The Arts Council of Great Britain, pp. 280–1; The Trustees of the British Museum, pp. 69, 75, 104–5, 127, 143, 152–3, 174, 197, 199, 200–1, 202, 203, 219, 225, 226, 233, 236–7, 239, 244, 248–9, 276–7, 319, 335, 347, 353, 356–7; British Transport Commission, pp. 164–5; Brown Lenox and Co., Ltd, p. 119; Camera Press Ltd, p. 89; J. Allan Cash, p. 77; The Cement and Concrete Association, pp. 43, 255, 301, 302, 306; Central Press Photos Ltd, pp. 209, 243; Gerti Deutsch, p. 71; Forth Road Bridge Joint Board, p. 249; German Archaeological Institute, Rome, p. 328; Martin Hürlimann, pp. 220, 221, 268; Husband and Co., p. 298; The Trustees of the Imperial War Museum, p. 147; The Institution of Civil Engineers, pp. 61, 235, 366; The Institution of Mechanical Engineers, pp. 113, 279, 288; The Mansell Collection, pp. 135, 312; Eric de Maré, p. 230; Georgina Masson, pp. 19, 305; D. C. Milne, M.I.C.E., pp. 364, 365; Admiral of the Fleet the Earl Mountbatten of Burma, pp. 280–1; The Trustees of the National Maritime Museum, p. 173; The Trustees of the National Portrait Gallery, p. 141; O. Margaret Pannell, p. 265; The Port of London Authority, pp. 151, 155; Radio Times Hulton Picture Library, pp. 82, 83, 88, 124, 125, 194, 247, 297, 327; The Science Museum, London, pp. 95 (Crown Copyright), 97, 101, 285 (Crown Copyright), 295, 312–13 (Crown Copyright), 317, 332 (Crown Copyright), 337; Edwin Smith, pp. 33, 175, 311; Colin Sorensen, pp. 213, 267, 359; Southampton Corporation, Museums Dept, p. 193; Stewart and Lloyds Ltd, p. 367; Swedish Tourist Board, p. 81; The Trustees of the Tate Gallery, pp. 129, 157; S. R. Thomason, p. 59; United States Information Service, pp. 252, 253; Jan Versnel, pp. 354–5.